PLANTS AND PEOPLE

Vegetation Change in North America

Thomas R. Vale
Department of Geography
University of Wisconsin
Madison

RESOURCE PUBLICATIONS
IN GEOGRAPHY

Library of Congress Cataloging in Publication Data

Vale, Thomas R.
    Plants and People

    (Resource publications in geography)
    Bibliography: p.
    1. Botany — North America — Ecology. 2. Vegetation
dynamics — North America. 3. Man — Influence on nature —
North America. I. Title. II. Series.
QK110.V34   1982   581.5'264'097   82-8865
ISBN 0-89291-151-4   AACR2

Publication Supported by the A.A.G.

Graphic Design by D. Sue Jones and CGK

Printed by Commercial Printing Inc.
State College, Pennsylvania

Cover Maps: Cartographic Laboratory, University of Wisconsin – Madison

Foreword

The "wild" vegetation around us reflects, in fact, the impact of people on plant communities. Natural and wilderness areas would not be so important were the effects of humans not so ubiquitous. From academic studies in plant geography and ecology to applied research in environmental impact assessment, range management, and logging practices, the literature on people-plant interactions is voluminous. Traditional views of plant succession have been questioned in the last several decades, with the emergence of new viewpoints on plant-environment equilibria. Similarly, such long-hallowed conservation practices as forest fire prevention have been challenged. "Smokey the Bear" may not have always been right!

It is appropriate that a review of human impacts on plants provide both a guide to the literature and a perspective from which vegetation change can be veiwed. Tom Vale has done both in this volume. By comparing human disturbances of vegetation to natural phenomena, Vale puts people in their ecological place. Our impacts are widespread and significant, but *ethics* and *values* rather than *science* must determine whether we are any different from natural disturbances in kind or consequence.

Resource Publications in Geography are monographs sponsored by the Association of American Geographers, a professional organization whose purpose is to advance studies in geography and to encourage the application of geographic research in education, government, and business. The series brings contemporary research in the various fields of geography to the attention of students and senior geographers, as well as to researchers in related fields. The ideas presented, of course, are those of the author and do not imply AAG endorsement.

In a volume of this scope, Vale could not hope to describe all the ecological roots of the various studies cited. Readers are encouraged to use the references in cited materials to explore the development of the ideas presented, and to use the scientific citation and abstracting publications to bring each topic up to date.

We are particularly grateful to the Cartographic Laboratory of the University of Wisconsin—Madison for the preparation of the figures in this monograph.

C. Gregory Knight, *The Pennsylvania State University*
Editor

Resource Publications Advisory Board

George, W. Carey, *Rutgers University*
James S. Gardner, *University of Waterloo*
Charles M. Good, Jr., *Virginia Polytechnic Institute and State University*
Risa I. Palm, *University of Colorado*
Thomas J. Wilbanks, *Oak Ridge National Laboratories*

Preface and Acknowledgements

I have been interested in human impacts on vegetation since my undergraduate years studying ecology at Berkeley. This interest was expressed indirectly in my Master's thesis on the Redwood National Park controversy, more specifically in parts of my dissertation on sagebrush as a landscape element, and explicitly in my subsequent research on tree invasions of meadows and brushfields. From these particular instances of vegetation altered by people, I found myself wondering about more general questions of vegetation change and the relationship between natural and human causes of plant disturbance. This book is the result.

At Wisconsin, I teach a semester course which treats the themes presented in the following pages. This book might be useful, then, as an outline for others wishing to develop such a course, or as a text for students. Moreover, it might also be used as a supplemental reference for a more general course in biogeography or related fields.

I am troubled by possible misinterpretations of the conclusions of this book. In the following pages, I stress apparent similarities between natural and human causes of vegetation change and emphasize several perspectives from which change may be viewed. I suggest that it is possible to equate the effects of drastic human alteration of plant cover with the effects of equally drastic natural changes. Because I identify such parallels, some readers may assume that I have failed to recognize the importance of the great contemporary vegetation transformations that are receiving so much attention — tropical deforestation, "desertification," and species extinctions. Nothing is further from the truth. However, I have been struck repeatedly, over fifteen years of looking at resource issues, that objective evaluations of such problems, whether economic or ecologic, cannot provide us with answers. We must ask ourselves if the results of our decisions will create a world with which we will be happy, as well as one which will suit our economic or ecologic needs. Our evaluation of human impacts on the plant cover of the Earth are no different from other resource problems. Vegetation is changeable. Some changes do, indeed, reduce the ability of our planet to support human life, but the condemnation of such changes comes more from our ethics than from our science. We survive because we *want* to survive. Moreover, many vegetation changes do not threaten our biological existence, but that fact does not make those changes any less important to our purposes as human beings interested in more than simple physical survival. I am impressed that whether we condone vegetation change or condemn it, we speak with our emotions about a vision of a desirable world. Everyone, whether business-person or bird-watcher, miner or Malthusian, economist or ecologist, evaluates vegetation change with a heart as much as with a head.

Several acknowledgements are appropriate. The editor of the AAG Resource Publications in Geography, C. Gregory Knight, was a supporter of this book from its inception and has given it much of his time and expertise in a variety of ways since. Anonymous reviewers of the initial prospectus and the draft were also helpful. A Ph.D. student at the University of Wisconsin, John Metz, kindly read the manuscript and made some useful suggestions. My wife, Gerry, not only aided me in thinking through many of the points discussed in the book but also put up with the curmudgeon that her husband too often became during its writing.

Thomas R. Vale

Contents

List of Figures

List of Tables

1

Introduction

Whenever a plant cover does not closely resemble either a cultivated field or a manicured lawn or park, it is often perceived as being "natural." Specifically, the forests of western mountains, the brushfields of the interior West, the grasslands of the High Plains, and the wooded hills of the eastern states are popularly thought to be largely free from alteration by modern society. In fact, most plant covers, and indeed most North American vegetation, have been changed since European contact by a variety of human impacts — altered fire regimes, logging, livestock grazing, trampling, off-road vehicle use, air pollution, changed biota, active manipulation of vegetation, construction activities, and abandonment of former farm land. A person amid vegetation that might seem to be entirely natural would most likely be surrounded by a plant cover that is a cultural artifact.

Two hypothetical examples illustrate the subtle effects of vegetation change. First, consider a wooded ridge in western Maryland (Figure 1a). In the valley below the ridge, frequent fires kept the vegetation open and grassy in precolonial times. After European settlement, the lowland was cultivated for crops before being abandoned. White pines and other trees have subsequently invaded the former fields. The massive white pines which originally grew at the top of the slope, in contrast, are gone, victims of logging. Although not illustrated in the figure, another woody species, pin cherry, has also decreased over the last century. Pin cherry is heavily browsed by white-tailed deer. Deer have increased greatly in numbers because of the abundance of food plants produced by timber cutting and elimination of burning. Finally, extending along the base of the slope is a narrow, linear strip of disturbed forest, the path of a recently buried natural gas pipeline. Someone may admire the setting as a natural scene, but it bears the imprint of many human activities.

Second, consider a brushy slope in central Nevada (Figure 1b). In the valley bottom, heavy livestock grazing has greatly reduced the grass and allowed an increase in rabbitbrush. Intense livestock use has also compacted the soil which, in turn, has increased surface water runoff and caused the stream to incise. On the slope, both brush and junipers have invaded from the rocky outcrop into the valley, in part because of grazing but also because of an elimination of occasional fires. Finally, an area of brush in the valley has been eliminated recently and planted to crested wheatgrass in order to improve forage for cattle. The rectangular planting pattern is the only clue that this landscape is anything but natural.

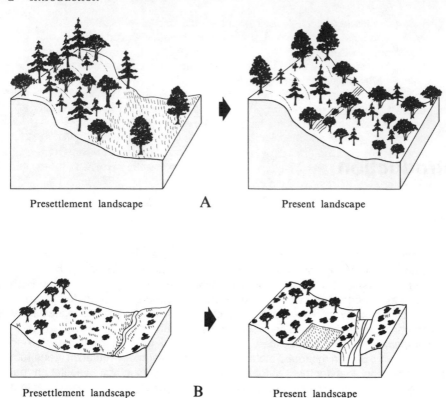

Presettlement landscape   **A**   Present landscape

Presettlement landscape   **B**   Present landscape

FIGURE 1   TWO HYPOTHETICAL LANDSCAPES ALTERED BY HUMAN ACTIVITIES. **A.** Ridge in western Maryland. **B.** Slope in central Nevada.

Factors Influencing Vegetation

These examples suggest the ubiquity of human impacts on vegetation. Human activities, however, contribute to, rather than substitute for, natural factors influencing vegetation characteristics and patterns. At least five such ecological factors can be identified:

(1) *Regional climate* determines the availability of water and energy, and thus is the dominant influence upon the general structure of vegetation.

(2) *Topography* modifies the availability of moisture by influencing local climate and water movement over the landscape.

(3) *The soil,* or *substrate,* may also affect moisture supplies, and it likewise is the major factor in the chemical relationships linking plants and environment.

(4) *Biotic* influences include both plant-animal and plant-plant interactions, the most common of the second being so-called competition.

(5) Finally, *disturbance events* such as fires alter local climate, substrate characteristics, and biotic interactions.

Taken together, these factors influence the structure of vegetation, including both the
arrangement of plant mass vertically and horizontally in space and the physical characteristics of that mass. In addition, the flora, or species composition of vegetation, is in
part determined by these same ecological factors. In the absence of humans, the five
environmental influences interact to produce a hypothetical plant cover, called potential
natural vegetation. More realistically, they combine with human impacts to create the
"natural" vegetation existing in the landscapes of North America (Figure 2).

Disturbance as a Vegetation Factor

The role of disturbance deserves special mention. Disturbance may be defined
with two characteristics: (1) It is an environmental change or event that makes plant
resources available (such as water or light) that were formerly fully utilized by the

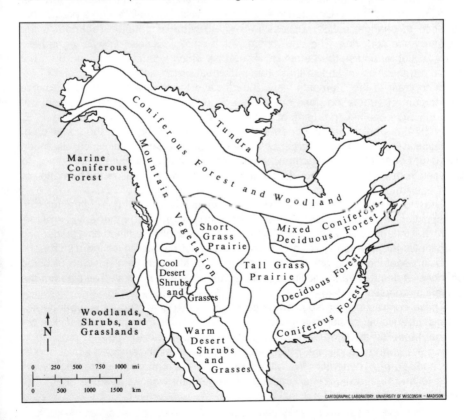

FIGURE 2   NATURAL VEGETATION OF NORTH AMERICA
(Bailey 1976)

pre-disturbance vegetation. (2) As a result of the freeing of the resources, vegetation change is initiated.

Many types of environmental change fit the description clearly, such as fire that creates an open and sunny forest floor, windstorms that blow down forests, or landslides that expose mineral soil. Other events may seem less clearly to be disturbance even though the effects conform to the definition given. For example, an increasing number of browsing animals may consume a favored food plant and eliminate it from a forest. Another less-palatable plant species may then increase in importance because of the availability of moisture or sunlight. A second example would be a drought that reduces a moisture-demanding plant, which, in turn, allows a drought-resistant species to increase. Are variations in animal numbers or weather conditions "disturbance," or merely "environmental fluctuation?" Our purposes in this book do not require answering such a question. We must, however, note that many definitions that attempt to encompass so complex a set of phenomena as ecological dynamics become fuzzy if looked at closely.

The role of disturbance has been a key issue in differing viewpoints about the nature of vegetation and vegetation change. Three models can be identified. Prior to about 1950, the conceptions of F. E. Clements dominated American ecology. Clements (1916) saw disturbance as an aberation to equilibrium vegetation that results from other environmental factors. A disturbance event was envisioned as initiating a predictable sequence of change, called succession, by which the vegetation returned to the predisturbance condition. The equilibrium which ended succession was termed the climax, a stable, undisturbed state of vegetation which was characterized by great internal organization much like that of an individual organism.

In contrast to the Clements view, others have found that the post-disturbance development of vegetation does not lead back to the predisturbance condition, but rather toward some new grouping of species and perhaps vegetation structure. F. E. Egler (1954) is particularly prominent in this regard. He argued that the condition of vegetation at the time of the disturbance was the most important factor which influenced subsequent vegetation development. Moreover, this condition varied from place to place and from time to time. Each disturbance event, then, would allow the formation of a new equilibrium.

The two views of disturbance as a kind of deviant influence can be contrasted with a perspective which places disturbance phenomena on par with other environmental factors that influence vegetation. A. S. Watt (1947) is usually identified as a pioneering advocate for this viewpoint. He documented cycles of change involving disturbance events in vegetation that otherwise did not change through time. Instead of being described as destructive of equilibria, then, disturbance was portrayed as a *part of* the equilibrium condition.

These contrasting and sometimes conflicting viewpoints about disturbance and post-disturbance vegetation change form a thread that runs through this book. We will find that these different ways of looking at vegetation, and the related problem of defining equilibrium conditions, reflect, in part, differences in temporal scale, spatial scale, and degree of generality. We will first explore these matters in natural vegetation, before turning to specific human impacts on vegetation where the same alternative viewpoints again arise.

2

Vegetation Change

Vegetation is always changing. From one minute to another a leaf may fall from a branch to the ground, a bud may be eaten by a grouse, or the rate of water loss from a blade of grass may increase. From one hour to another, flowers may open, fire may sweep through a patch of brush, or net energy storage by photosynthesis may replace net energy loss by respiration. From one day to another, twigs may elongate with spring growth, freezing rain may cause breakage of tree limbs, or herbaceous perennials may push up through melting snow. From one month to another, a deciduous forest may become leafless with the onset of cold winter weather, leaf-eating insects may consume the foliage of trees and shrubs, or an annual grassland may turn from growing green to dormant golden. From one year to another, a large seed crop may be followed by a small one, a fire-consumed woodland may be replaced by a field of flowers and grasses, or the lush growth of one spring may be contrasted with the withered growth of the following. From one decade to another, an open meadow may become filled with invasive shrubs, a forest destroyed by a windstorm may recover with young trees, or the composition of a grassland may change as a wet period is followed by drought. From one century to another, a forest freed from burning may shift from fire-dependent pines to fire-sensitive maples; a timberline boundary at high latitudes may shift poleward during a warming trends; or a formerly abundant tree may be nearly eliminated by an unexplained increase in the numbers of a browsing animal. Over still longer time scales, other factors may contribute to a constantly changing plant cover — migration of plant species, changes in climate, evolutionary change in the genetics of species, or progressive changes in soil characteristics.

We cannot review in this book all of these types of change, nor should we. Our purpose is more restricted — we want an overview of those changes that will provide us with a perspective from which we can evaluate human impacts on vegetation. Two characteristics of change seem particularly appropriate.

First, our concern is focused primarily on intermediate time scales. That is, most human activities that alter plant communities are most properly compared with other causes of change that are discernible over a few decades, a century, or several centuries. Change which occurs over either a few minutes or a few millennia is less central to our purpose, although even those scales might be helpful in some situations. Thus, our evaluation of logging, for example, will focus on vegetation change over the 10, 50, or 100 years after timber removal, and not so much either on the hours during which the trees are cut or on the thousands of years which may follow a single logging event.

Second, we are largely concerned with changes that we can identify as disturbance. That is, most human activities seem to alter the natural disturbance processes of vegetation. Changes in disturbance regimes may include the elimination of a disturbance agent (suppression of fires), the introduction of a new disturbance agent (trampling), or the substitution of one disturbance agent for another (substituting cattle for bison).

Taken together, these two characteristics of change — time and disturbance — may be grouped in a two-part classification of human impacts on vegetation. First, human activities may involve low or moderate levels of persistant disturbance recognizeable over several decades or a century or two. Examples are livestock grazing, trampling, off-road vehicle use, and air pollution. Second, human activities may represent intense disturbance of short duration. In such cases, the development of the vegetation over several decades or even a century after disturbance provides the primary measure of the human impact. Examples are fires (or development of vegetation with fire suppression), logging, construction activities, and farm abandonment.

Comparisons between natural and human-induced vegetation change should focus, then, on intermediate time scales (but include other time scales where they help give perspective), and on disturbance as a factor initiating change. Both of these themes are apparent in the first example of natural vegetation change which follows — change which maintains a given state of the plant cover through cycles of reproduction.

Vegetation Change Around an Equilibrium

Certain types of vegetation change may be viewed as variations around some sort of average or mean condition. These cyclical changes involve an equilibrium which persists through time. Disturbance events are, from this perspective, a part of the vegetation system, which itself fluctuates but maintains its essential characteristics.

Replacement Processes

Vegetation dominated by species with characteristics which allow indefinite self-perpetuation on a site often undergoes considerable change through time. Consider, for example, a small pure stand of sugar maple *(Acer saccharum)*, a tree which may occupy moist sites in middle-latitude environments indefinitely in the absence of disturbance. When mature canopy trees die, light and moisture increase on the forest floor and large numbers of sugar maple seedlings survive. As the new seedlings grow to saplings, the dense growth reduces light and water for subsequent seedlings, and new reproduction is greatly reduced or entirely precluded. Not until the saplings have matured and at least some die will another new generation become established. Most reproduction thus occurs on the site only periodically, rather than continuously. The forest as a whole consists of stands, or "patches," at differing stages within this cycle. Waves of reproduction have been identified for the classic climax species, sugar maple (Bray 1956) and eastern hemlock *(Tsuga canadensis)* in North America (Hett and Loucks 1976). Earlier work by Watt (1947) documented such a replacement cycle for beech *(Fagus sylvatica)* in western Europe.

Reproduction cycles may depend upon modest break-ups of the canopy of a mature forest. Other replacement cycles, however, involve localized disturbance by more catastrophic events. For example, some ponderosa pine (western yellow pine;

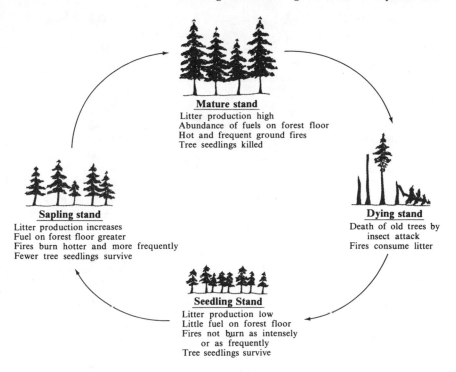

Mature stand
Litter production high
Abundance of fuels on forest floor
Hot and frequent ground fires
Tree seedlings killed

Sapling stand
Litter production increases
Fuel on forest floor greater
Fires burn hotter and more frequently
Fewer tree seedlings survive

Dying stand
Death of old trees by
insect attack
Fires consume litter

Seedling Stand
Litter production low
Little fuel on forest floor
Fires not burn as intensely
or as frequently
Tree seedlings survive

**FIGURE 3   STAND REPLACEMENT CYCLE IN A
PONDEROSA PINE FOREST**

Pinus ponderosa) forests consist of patches of even-aged trees which become established after the death of stands of mature trees caused by insect attack or fire (Figure 3; Biswell 1973). More recently and dramatically, forests of balsam fir (Abies balsamea) in the mountains of New England have been shown to reproduce in narrow but elongated strips where high winds have caused the death of canopy trees. The young trees survive into maturity until winds again break up the canopy and allow reproduction (Sprugel 1976; Sprugel and Bormann 1981). Reiners and Lang (1979) have identified other disturbance agents in the balsam fir forests (avalanches, hurricanes) which, together with strips of wind-induced reproduction, produce a mosaic of stands of varying ages and sizes. A more simple pattern is produced in the alpine environments of Colorado where patches of trees may experience mortality on their windward sides and successful vegetative reproduction on their leeward edges. Areas where new individuals become established will thus progress through time into a stand of mature trees before the patch degenerates and is replaced by non-forest vegetation (Marr 1977).

Each of the replacement cycles discussed has portrayed periodic reproduction by trees without involving other species. Replacement cycles by which certain species replace themselves after other intermediate species have occupied the site may be just as common. Watt (1947) identified the classic cycles of replacement involving heather (Calluna vulgaris), bearberry (Arctostaphylos ura-ursi), and lichen (Cladonia silvatica)

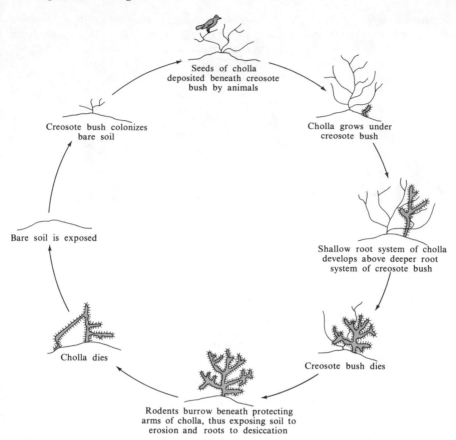

FIGURE 4   REPLACEMENT CYCLE INVOLVING CREOSOTE BUSH AND CHRISTMAS TREE CHOLLA IN TEXAS (Yeaton 1978; reproduced by permission of Blackwell Scientific Publications Limited).

on heath lands. He also outlined a sequence on peatlands in which mosses invade shallow pools and build up accumulations of plant material to form mounds. These hummocks are invaded by heather which matures and eventually dies, thereby exposing the soil to erosion and reformation of the pools. Billings and Mooney (1959) described a similar cycle of peat accumulation and subsequent erosion on alpine tundra in Wyoming. In addition, Yeaton (1978) found a replacement cycle in which creosote bush *(Larrea tridentata)* is replaced by cholla cactus *(Opuntia leptacaulis)* which in turn is replaced by creosote bush (Figure 4) in the deserts of northern Mexico and the American southwest.

In the eastern forests of the United States, species replacement cycles are common (Stephens 1956). Tree species requiring high light intensities become established when a hole, or gap, in the canopy develops, but through time these trees are replaced by shade-tolerant species. With new gaps being created continuously at differing places in a forest, such "gap-phase" phenomena may allow light-demanding trees to perpetuate themselves indefinitely even if the forest is not subjected to catastrophic disturbance. Forcier (1975) has argued that a sequence of yellow birch *(Betula alleghaniensis)*, sugar maple, and American beech *(Fagus grandiflora)* may be repeated at a given site within a mesic forest (Figure 5). In invoking the general gap-phase model, Bormann and Likens (1979) stress the important interplay between the age of the canopy trees (and thus their susceptibility to death by windthrow), the size of the gap produced, and the varying ways in which species may exploit the opening-colonization of the site by seed migration, rapid growth into the canopy of established but long-suppressed individuals, or germination of seed previously buried on the site (Table 1.)

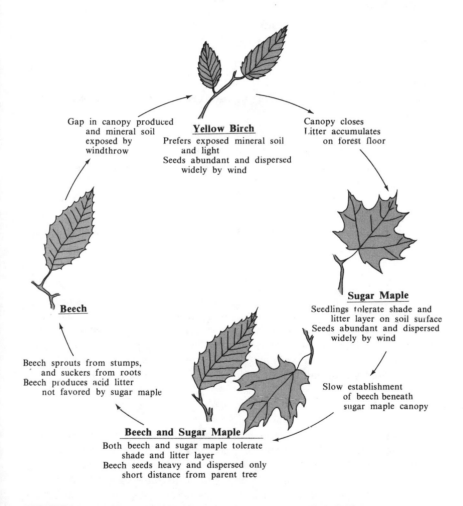

FIGURE 5 REPLACEMENT CYCLE IN A YELLOW BIRCH-SUGAR MAPLE-BEECH FOREST (Forcier 1975; reproduced by permission of L. K. Forcier and *Science*. Copyright 1975 by the American Association for the Advancement of Science).

TABLE 1   STRATEGIES BY WHICH PLANTS REOCCUPY FOREST OPENINGS CREATED BY DISTURBANCE EVENTS

I. **Outgrowth.** Lateral or horizontal expansion of established individual plants. Most important strategy for small openings.

    A. Clonal expansion of herbs and shrubs

    B. Expansion of crowns and roots of trees surrounding openings

II. **Upgrowth.** Upward growth by plants from forest floor. Small openings in canopy may involve shade-tolerant species, larger openings may be dominated by species intermediate in shade tolerance, and still larger openings provide the greatest opportunities for shade-intolerant species.

    A. Small/young trees on forest floor prior to creation of opening ("Advance regeneration")

        1. Sprouts from roots or bases of trunks

        2. Seedlings or saplings

    B. New individuals

        1. Sprouts from roots or stumps of established trees injured by disturbance event

        2. Seedlings

            a. Seeds buried or stored from years prior to disturbance event

            b. Seeds produced in year of disturbance event

                1) Heavy seeds produced immediately before or soon after disturbance event by trees within or adjacent to forest opening.

                2) Light seeds produced either near forest opening or far from opening and dispersed to site by wind

Source: Bormann and Likens 1979: 105 and 132. Reproduced by permission of Springer-Verlag New York.

All of the replacement processes described, in fact, involve strong interactions between the condition of the pre-disturbance vegetation and the effects of disturbance events. For example, in the sugar maple cycle, mature trees, not young trees, are most susceptible to windthrow. Moreoever, in the cycle involving creosote bush and cholla, the cactus becomes vulnerable when it has grown sufficiently large to act as protection for burrowing rodents. These kinds of interactions suggest that disturbance events and their effects are inescapably tied to the vegetation which is being disturbed.

Succession

Replacement processes typically involve localized disturbance, which initiates changes that may not be discernible when considering large areas of landscape. Disturbances may occur at various scales, however. If a particular event is effective over a large area, "secondary succession" may occur. Classically, secondary succession is seen as a gradual reestablishment of the predisturbance vegetation, with certain trends during recovery being similar regardless of the species or vegetation involved. Major catastrophic disturbance will generally favor shade-intolerant species and often herbaceous plants of low stature. The post-disturbance development of the vegetation will result in an increase in shade-tolerant species, although intolerant species will normally persist because of continued smaller-scale disturbance such as gap-phase phenomena. Non-woody plants decline in importance in favor of woody species, and the height of the plant cover increases. Odum (1969) has argued, in fact, that many characteristics of ecosystems, including species diversity, niche specialization, and mineral cycling, change in predictable ways during succession. Horn (1974), in contrast, has stressed the variability in changes during succession, and he has pointed out that much of the conventional wisdom regarding secondary succession is incorrect. For example, diversity does not increase regularly with sucession. Rather, "some intermediate stage should have a mixture of both early and late species, and thus a higher diversity than early or late stages" (Horn 1974:31). In addition, dynamic stability (the ability to rebound after disturbance) does not increase with succession, but actually decreases. The reasoning is simple: "Disturb early succession and it becomes early succession. Disturb a climax community and it becomes an early successional stage that takes a long time to return to climax" (Horn 1974:32). Finally, productivity does not achieve maximum values in the "climax", but it actually decreases toward the "end" of the successional sequence. It is not surprising that the concept of succession has received considerable evaluation and criticism through the years (Drury and Nisbet 1973; McIntosh 1979).

In situations where the return time for events causing widespread disturbance is short relative to the rate of vegetation change, the plant cover may be maintained indefinitely in an "early successional" state. The absolute times involved may vary considerably. For example, in the precolonial tall grass prairie, fires probably burned most years, maintaining vegetation free of woody plants (Bragg and Hulbert 1976). Mixed coniferous forests of the Sierra Nevada were burned once every decade or two, and thus the shade-intolerant pines were maintained as dominant elements in the forest (Kilgore and Taylor 1979). In the Valdivian Andes of Chile, landslides that have occurred once every century or so have retarded the successional development of the forests and have helped maintain shade-intolerant tree species (Veblen and Ashton 1978; Veblen 1981). In situations such as these, ecologists today usually include the disturbance regime as a component of the environmental system of which the vegetation is a part (Drury and Nisbet 1971; Veblen and Ashton 1978). The idea of a predictable secondary succession leading toward a self-perpetuating "climax" condition is not as useful as often thought. It seems to add little to an understanding of the forces which influence vegetation characteristics and patterns at a particular place and time.

Environmental Fluctuations

Temporal changes in environmental conditions may cause corresponding alterations in vegetation which persist until the environment — and thus the vegetation — reverts to the earlier state. Such environmental fluctuations may be grouped with fire and windthrow as disturbance events, or they may be segregated in a distinctive category of changing environmental characteristics. In either case, the underlying concept of cyclical change in vegetation remains valid.

Climatic fluctuations, for example, may influence the survival of established plants or alter the species mix of successful reproduction. During a warm, dry period, a grassland may experience altered composition. When more moist conditions return, the grassland may change back to its earlier makeup. In addition, animals often influence the characteristics of vegetation, so that fluctuations in animal numbers may result in short-lived variations in plant cover. Finally, seeds produced by plants are seldom equally abundant each year. Therefore, the rate of reproduction may also vary through time. In each of these examples, it is assumed that the source of novelty in the environment, whether climate, animal numbers, seed production, or some other variable, is varying around some sort of mean, and eventually the environmental conditions will return to earlier characteristics.

Disturbance Around an Equilibrium

Various types of disturbance phenomena around an equilibrium may be schematically represented in a single diagram (Figure 6). Each box represents a 'state' of the vegetation, although in reality any 'state' is a 'snap-shot' segment of vegetation that in fact varies from place to place in the landscape ("spatial variation") and from time to time at one spot ("temporal variation"). The various changes in vegetation are represented by arrows which indicate the alteration of the plant cover from one 'state' to another. Underlying this scheme is the assumption that the vegetation has some sort of equilibrium or balance with environmental conditions.

Vegetation Change with New Equilibria

Some vegetation changes may be viewed as creating new equilibrium conditions. Such change is termed "developmental," to distinguish it from cyclical fluctuations that perpetuate established equilibria. Disturbance events that initiate developmental change are, in this perspective, the means by which new states of the vegetation are created.

Short-Term Developmental Change

In contrast to alternating episodes of disturbance and subsequent vegetation development, disturbance events may be also envisioned as allowing the formation of new equilibria. Frequently changing equilibrium conditions result from numerous factors which influence the development of vegetation after disturbance and which make the post-disturbance recovery of the plant cover variable. Several factors promote this variability:

(1) Environmental conditions at various sites within a disturbed area are often different. Post-fire vegetation development on a warm and dry south-facing slope, for example, may be different from that on a cool and moist valley bottom.
(2) Environmental conditions at the same site may vary from time to time. For example, year-to-year fluctuations in weather may alter the mix of seeds being produced by species within vegetation, thereby influencing the availability of seeds when disturbance occurs.
(3) Conditions produced by various disturbance agents are different and thus may alter the composition of the post-disturbance vegetation. For example, fire may consume litter on the forest floor and encourage tree species which require a mineral seedbed for

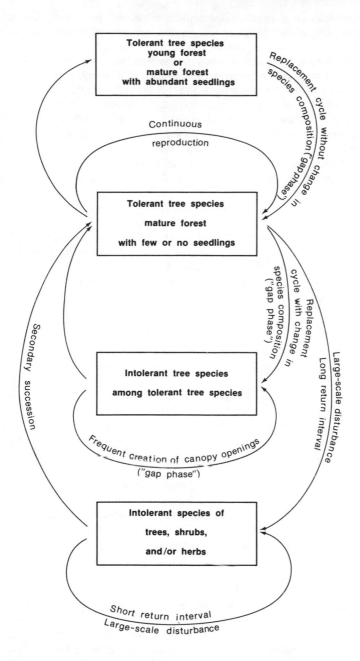

FIGURE 6 SUMMARY OF NATURAL CHANGE IN FOREST VEGE-
TATION AROUND AN EQUILIBRIUM. The boxes represent vegetation
states. Arrows from one box to another represent changes in vegetation
conditions, and the arrows which start and end at the same box represent
processes which maintain vegetation conditions.

reproduction, whereas major blow-downs of large trees may produce an organically-rich seedbed which encourages different tree
species.

(4) The array of species which are able to take advantage of the
resources made available by disturbance vary both spatially and
temporally, a factor stressed by Egler (1954). For example,
shrubs and herbaceous plants may dominate a forest site after
disturbance if they are present in the pre-disturbance vegetation.
If they are not present, tree seedlings may become established
immediately after the disturbance.

The composition of the pre-disturbance vegetation as well as the composition of nearby
undisturbed vegetation are both important influences which contribute to variable
post-disturbance circumstances.

Several examples illustrate short-term developmental change. Henry and Swan
(1974) argued that over 300 years, a New Hampshire forest experienced compositional
change coincident with major disturbance events (fire, windthrow), but that little of the
change could be described as cyclical or successional. Rather, the changes in the
forest reflected the capabilities of particular species to take advantage of the conditions
created by the disturbance events. Similarly, Oliver and Stephens (1977) found that
additions to the canopy of a forest in Massachusetts occurred mostly at times of
disturbance, either by the establishment of new individuals or by the growth of smaller
trees into the canopy. Successional replacement was not an important process. Oliver
(1981) later argued that most North American forests in precolonial times were dominated by individual trees established immediately after disturbance events rather
than by trees gradually replacing one another through time. Also, Auclair and Cottam
(1971) have suggested that the composition of woodlots in Wisconsin reflects the
establishment of tree species during periods of adjustment to new environmental
conditions (climatic drought, livestock grazing), but that the changes could not be
envisioned as cycles of successional replacement. In addition, studies of vegetation
development suggest that the composition of forests after fires varies from site to site,
depending on the availability of seeds of different species (Lyon and Stickney 1976;
Cattelino et al. 1979; Antos and Habeck 1981).

Short-term developmental change incorporates several features: (1) equilibrium
conditions persist between disturbance events; (2) new equilibria are established with
each major disturbance event; and (3) successional sequences do not return the
vegetation to pre-disturbance conditions. The diagram portraying disturbance
phenomena must be altered accordingly (Figure 7).

Long-Term Developmental Change

Vegetation may experience developmental change over long time scales, usually
in response to environmental change. This environmental variability may be defined as
disturbance if one wishes to broaden the context appropriate to the definition, although
to do so is uncommon. The variability is more often classified as "environmental
change." Regardless, the developmental change of vegetation is what concerns us
here.

Sometimes long-term developmental change in vegetation is abrupt rather than
continuous. For example, the major climatic change represented by the Pleistocene-
Holocene boundary (the end of the Ice Age, about 10,000 years ago) was an abrupt

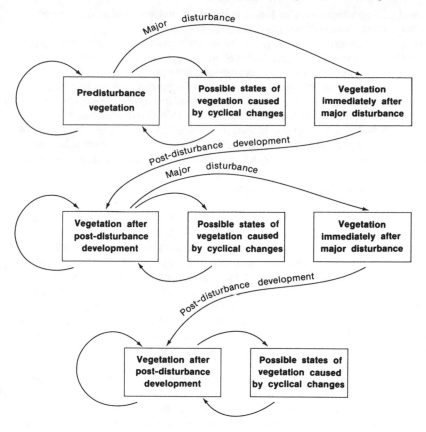

FIGURE 7 NATURAL CHANGE IN VEGETATION WITH CREATION OF NEW EQUILIBRIA AFTER MAJOR DISTURBANCE EVENTS

alteration in environmental conditions. The structure of the vegetation of North America, adjusting to the warmer climates, responded relatively rapidly. However, variable rates of species migrations caused the vegetation history of particular sites to reveal reoccurring abrupt changes in composition and perhaps in ecological dynamics with the arrival of new species (Davis 1976; Wright 1976). In addition, similar abrupt patterns of change may reflect the arrival or development of pathogens. Davis (1978), for example, attributed the widespread decline in hemlock in eastern North America about 5000 years ago to an outbreak of a microorganism or insect specific to that tree. With hemlock reduced, other tree species increased in importance, although the mix of species responding to the decreased competition with hemlock varied from place to place. Finally, abrupt changes may occur in relation to new vulnerability to disturbance phenomena. For example, the recent introduction of cheatgrass *(Bromus tectorum)* in the northern intermountain West has altered the fire regime of sagebrush-grass rangeland. Cheatgrass is an annual species that dries in the summer and carries hot fires much better than did the previously dominant bunch-grasses. With the increase in intensity and perhaps frequency of fire in that region, the vegetation is being transformed from a shrub/bunch-grass plant cover to annual grasses.

Long-term developmental change may involve gradual, rather than abrupt, changes in environmental conditions. Vegetation-soil systems may undergo progressive change at slow rates, discernible only with long time perspectives. The spodosol/ pygmy forest ecosystem on the northwest coast of California may represent such a system in which the vegetation and soil have developed gradually for perhaps millions of years (Jenny *et al.* 1969). The decrease in American elm *(Ulmus americana)* and other hardwoods from early Holocene forests of Minnesota has also been ascribed to a progressive leaching of glacial drift (Amundson and Wright 1979). Genetic changes in plant species may also be a source of gradual vegetation change at a long-time scale.

Human Activities and Vegetation Change

Many issues involving human impacts on vegetation are questions of whether human activities correspond to *equilibrium-maintaining* phenomena in natural vegetation, and thus are benign, or impose new equilibria and as *developmental* are potentially harmful to human interests. Several examples may be helpful. Clearcutting of forests is often attacked as an essentially destructive process in which a forest equilibrium is transformed into a non-forest equilibrium. However, defenders of clearcutting often argue that such logging practices merely replicate catastrophic disturbance events which occurred in precolonial time, and that the forest will return to its predisturbance condition in the same way that it has repeatedly in the past. As a second example, fire suppression may be interpreted as allowing the accumulation of fuels which thereby change the ecological behavior of fires. New equilibrium conditions involving hotter, less frequent fires replace those of cooler, more frequent burns. Conversely, elimination of wild fire in a fire-prone environment is impossible, one might argue. Thus, a fire-suppression policy cannot change fundamental vegetation features, and a pre-existing equilibrium condition may not be permanently changed by imposition of fire suppression. Construction activity in a grassland, such as strip-mining or laying a pipeline, might be viewed as drastic alteration of the pre-existing plant cover, particularly if the soil is transformed in the process. On the other hand, construction may be seen as equivalent to a natural disturbance, such as drought or heavy grazing by native animals.

In evaluating the effects of specific human activities on vegetation, various responses of plant communities to natural disturbance provide a context for evaluating the importance of disturbances caused by people. In addition, natural fluctuations in vegetation remind us that vegetation change cannot be considered apart from spatial and temporal scale. The importance of scale is particularly apparent in what might be the major human disturbance, changes in fire regimes.

3

Human Impacts on Vegetation

Against the background of natural processes of vegetation change, various human activities that effect plant communities can be assessed. Human disturbances include changing fire regimes, grazing, logging, trampling, off-road vehicle use, air pollution, construction, plant and animal introductions (and extinctions), agricultural land abandonment, and purposeful ecological manipulation. Clearly, the impacts of these disturbances will vary in continuity or abruptness, spatial and temporal scale, and intensity. Nevertheless, from an evaluation of these disturbance processes and their consequences we can derive some valid generalizations about human impacts on vegetation.

Changes in Fire Regimes

Most human activities that alter vegetation involve increases in disturbance. The most widespread human impact in North America, however, represents a *decrease* in the frequency of disturbance events: the suppression of wild fire. Fires have long been appreciated as an influence upon vegetation (Marsh 1864; Sauer 1950; Stewart 1956). The effects of burning are varied and numerous (Wright and Heinselman 1973; Kozlowski and Ahlgren 1974), and often involve dramatic changes in plant cover. Biomass, both living and dead, is usually reduced by burning, thereby releasing minerals to the soil (and perhaps substituting for biological decomposition in some environments). Fuels for subsequent fires are diminished. The composition of the vegetation is typically altered, both through differential survival of species (thick-barked trees survive better than thin-barked trees), and differential reproduction. Species requiring sunlight and mineral seed beds often are favored by burning. The post-fire development of vegetation, in fact, is often viewed as a successional sequence, but it may represent simply the growth of plants, present in the pre-fire vegetation, which are encouraged by burning (Lyon and Stickney 1976; Cattelino et al. 1979). Characteristics of the soil may be altered. The surface litter layer may be eliminated, or a layer which does not absorb water readily may form low in the profile. Hydrological characteristics of an environment may also be changed; streamflows may increase because of decreases in transpiration water loss. Loucks (1970) has suggested that periodic burning may be necessary for the maintainance of high productivity and high diversity in many ecosystems, and that species may evolve in response to the temporary conditions created by occasional burning. Moreover, vegetation tolerant of burning may have characteristics, such as more fats and flamable material in the foliage, which encourage the ignition and spread of fires. Species that are not favored by burning have characteristics that discourage easy ignition (Mutch 1970). Thus, vegetation does not simply respond to fire but actually influences the likelihood and spread of fire.

Precolonial fire regimes for different vegetation types in North America are typically determined by analyses of fire scars of living trees (McBride and Laven 1976; Arno and Sneck 1977) and occasionally by mapping the age structure of forests over large areas (Van Wagner 1978). In an environment lacking trees, knowledge about the fire ecology of the area's plant species may be compared with a description of vegetation characteristics, thus permitting an inference about the fire regime. The development of vegetation after recent fires (Bragg and Hulbert 1976; Shinn 1980) and early journal accounts or diaries (Curtis 1959; Barrett 1980) may be useful in these inferences.

Forest fires occurred at differing magnitudes and frequencies (Heinselman 1978). Some vegetation types, such as the boreal forest, were characterized by relatively infrequent, large-magnitude fires in which the canopy of the vegetation carried the flames, and the forest as a whole was largely killed. Such fires are called "crown fires" because they burn in the upper foliage or crowns of the trees. Other environments, such as lower-elevation, ponderosa pine forests in the western United States, had a regime of frequent, small-magnitude surface fires. Here, the burning was restricted to the forest floor, or "surface," and most mature trees survived. Variations of these two fundamental fire types also occurred, reflecting a continuum in both magnitude (light-surface, severe-surface, and crown fires) and frequency (short, intermediate, or long return times). Moreover, many forests had fires of different magnitudes recurring at different frequencies. Therefore, characterization of precolonial North American fire regimes is difficult. Not only have few vegetation covers been studied in detail, but also fire regimes, themselves, are often complex.

Even with these reservations in mind, generalizations about precolonial fire regimes in the United States and Canada are possible (Figures 8 and 9). Few vegetation types were free from recurrent fire. Tundra, alpine, and warm desert environments had too little fuel. Certain forests of New England (Siccama 1971; Lorimer 1977), forests in moist topographic situations (Arno and Davis 1980; Romme and Knight 1981), and perhaps, forests in the southern Appalachians (Lorimer, 1980) apparently were not strongly influenced by fire. Most other forested environments burned with some regularity, although the frequency was highly variable, and many were affected by both crown and surface fires. A regime of stand-replacing crown fires and few surface fires were most characteristic of the high latitude boreal forest and some high elevation forests in the West. In both lower latitude eastern states and lower elevations in western mountains, surface fires become more important. In southeastern pine forests and low elevation western ponderosa pine forests, a regime of frequent surface fires, and few, if any, crown fires, was characteristic. Areas with an abundance of herbaceous vegetation seemed to have fire regimes of frequent surface burns. This herbaceous vegetation had either no woody plants (as in the prairies), or scattered trees or shrubs (as in southeastern pine forests, desert grasslands, sagebrush-scrub, or grassy western coniferous forests).

In some situations, European settlement increased fire frequencies or intensities. For example, initial settlers may have been responsible for increased fire frequencies either by carelessness or by the desire to clear forests or encourage grass (Shinn 1980). The burning associated with early logging was often more intense than "natural" fires, because of larger amounts of fuels (Marquis 1975). An increase in the fire frequency and/or fire intensity in the sagebrush grass vegetation of the northern intermountain West has been associated with the introduction of the Eurasian cheat grass (Young and Evans 1978).

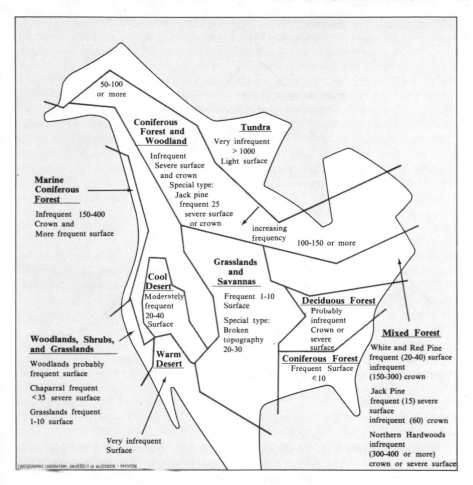

FIGURE 8   PRECOLONIAL FIRE REGIMES IN LOWLAND NORTH AMERICA. Frequencies are expressed in years between fires and verbally, as are intensities. The characterizations are from published studies of fire regimes or from my estimates of the fire regimes based on the fire ecology of the plant species in the areas mapped (Ahlgren 1974; Arno 1976; Arno 1980; Arno and Davis 1980; Barney and Frischknecht 1974; Biswell 1973; Bonnicksen and Stone 1981; Bragg and Hulbert 1976; Burkhardt and Tisdale 1976; Cole 1977; Cole 1981; Cooper 1960; Curtis 1959; Davis et al. 1980; Dodge 1975; Habeck and Mutch 1973; Hall 1980; Hall and Homesley 1966; Harniss and Murray 1973; Heinselmann 1978; Houston 1973; Johannessen et al. 1971; Kilgore and Taylor 1979; Komarek 1974; Loope and Gruell 1973; Kozlowski and Ahlgren 1974; Lorimer 1977; Lorimer 1980; Martin et al. 1976; McBride 1974; McNeil and Zobel 1980; Parsons 1976; Parsons 1981; Sauer 1950; Shinn 1980; Siccama 1971; Stokes and Dieterich 1980; Taylor 1973; Vale 1979; Wagener 1961; Weaver 1961; Wells 1965; Wright 1980; Wright et al. 1979; Wright and Bailey 1980).

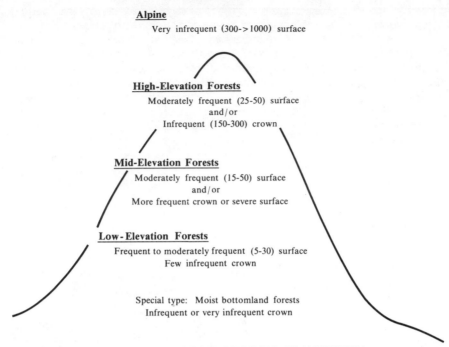

Alpine
Very infrequent (300->1000) surface

High-Elevation Forests
Moderately frequent (25-50) surface
and/or
Infrequent (150-300) crown

Mid-Elevation Forests
Moderately frequent (15-50) surface
and/or
More frequent crown or severe surface

Low-Elevation Forests
Frequent to moderately frequent (5-30) surface
Few infrequent crown

Special type: Moist bottomland forests
Infrequent or very infrequent crown

FIGURE 9   PRECOLONIAL FIRE REGIMES IN WESTERN MOUNTAINS OF NORTH AMERICA (See caption for Figure 8).

Fire Suppression

Although increased fire frequency and/or intensity occurred in some places, a reduction in burning has been most characteristic of European settlement. The impacts of this reduced incidence of fire follow from our discussion of fire ecology.

Increases in woody plants. In grassland environments protected from fire, invasion of woody plants is well-documented. Both in the earliest European settlement of the tall-grass prairie (Sauer 1950; Curtis 1959) and in contemporary times where fire has been eliminated from still-existing prairie (Bragg and Hulbert 1976), trees and shrubs increased greatly with fire exclusion. The grasslands of the Great Plains may owe some of their maintenance to fire (Wells 1965), although rapid expansion of woody plants may be precluded by the relatively dry climate (Borchert 1950; Muller 1971) and the lack of seed sources. In southwestern desert grasslands, the elimination of burning at least contributed to the post-Hispanic spread of trees and shrubs (Wright 1980; Wright and Bailey 1980). In the sagebrush-grass vegetation of the northern intermountain West, fire suppression may have been a factor in the decline of grass and the increase in shrubs (Harniss and Murray 1973).

Changes in forest composition. The elimination of fire caused a change in tree species which are reproducing in many forests. Burning encouraged trees that tolerated or preferred disturbance. Fire suppression, in contrast, has permitted shade-tolerant trees to germinate, survive, and dominate the seedling layer and understory (Parsons and De Benedetti 1979; McNeil and Zobel 1980; Parker 1980; Cole 1981).

The size and age structures of forests in a wide variety of locales display the resulting dichotomy between the fire-dependent, non-reproducing overstory inherited from pre-colonial times, and the fire-sensitive understory which has developed more recently (Figure 10). Such a pattern is most characteristic of forests that had frequent surface burns, but similar interpretations have been made of forests typified by crown fires (Day 1972; Habeck and Mutch 1973). In addition, other compositional changes in woody vegetation may result from altered fire regimes. For example, stands of quaking aspen *(Populus tremuloides)* do not reproduce well in the absence of fire (Gruell and Loope 1974), while stands of the intermountain shrub bitterbrush *(Purshia tridentata)* are reduced by burning (Wagstaff 1980).

Increases in forest density. Ground fires helped maintain open forests by killing much of the understory. Elimination of frequent burning has allowed the survival of seedlings and saplings, thereby increasing the density of forests (Bonnicksen and Stone 1981; Kilgore and Taylor 1979).

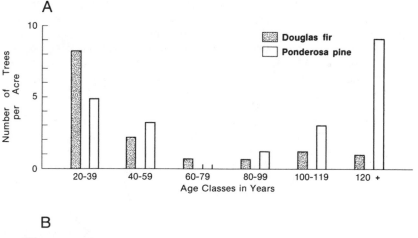

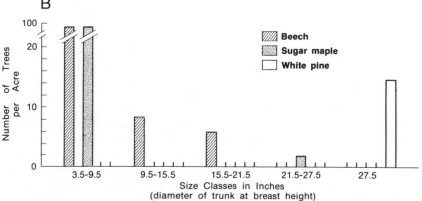

FIGURE 10 FOREST STRUCTURES EXPERIENCING COMPOSITIONAL CHANGES CAUSED BY FIRE SUPPRESSION. **A.** Oregon forest (Goff 1967). **B.** Northern Wisconsin forest (Dickman 1978, reproduced by permission of Society of American Foresters).

Changes in fire characteristics. The increases in forest density have made possible the spread of surface fires into forest canopies. Areas that formerly were characterized by frequent ground fires have now become susceptible to more catastrophic crown burns (Kilgore and Taylor 1979; Hall 1980). Moreover, the absence of occasional fires in areas that had crown fires has similarly permitted accumulations of fuels. Thus, subsequent fires are likely to be much hotter and more destructive than those of the precolonial regime. This change toward more intense fires seems also to apply to brush vegetation (Parsons 1976; Rundel and Parsons 1979; Bonnicksen and Lee 1979) and to drained wetlands (Taylor 1980), as well as to forests. The greater intensity of contemporary fires may explain why recent, but not earlier, fires killed individuals and restricted distributions of some species such as Coulter pine *(Pinus coulteri;* Vale 1979).

Changes in mosaic character of vegetation. Increases in fire severity caused by the accumulation of fuels has resulted in decreases in the spatial variety of some vegetation. Frequent fires burning through California chaparral stands burned more intensely in some places than in others, and thereby created a mosaic of stands of different characteristics. Subsequent fires then burned hottest in areas not burned earlier, but might not ignite stands recently burned. Fire suppression allows the widespread accumulation of fuels so that chaparral fires today tend to be uniformly intense over large areas (Parsons 1976). Similar mechanisms probably have also reduced the "patchiness" of forest vegetation (Arno and Davis 1980; Bonnicksen and Stone 1981).

Ectotone movements. Some major boundaries or ecotones between plant covers of differing structure are fire-related. The elimination or reduction in frequency of fires may initiate a movement of fire-sensitive plants into areas where burning previously precluded their survival. The movement of woody plants into grasslands is an example. The common invasion of sagebrush-grass vegetation by junipers may also be related to reduced fire frequency (Burkhardt and Tisdale 1976). Tree invasion of meadows in western mountains may sometimes be related to elimination of meadow grass burning (Gibbens and Heady 1965; De Benedetti and Parsons 1979).

Changes in Equilibria

Many effects of reduced burning are considered undesirable. As a management tool, fire is widely advocated in range environments where expansion of woody plants has reduced economically useful grasses (Pechanec 1965; Biswell 1977; Wright and Bailey 1980). In commercial forestlands, fire is still not commonly used, except in the southeastern pine forests where burning has long been utilized to encourage pine reproduction (Komarek 1974). Reintroduction of fires has been particularly newsworthy in landscape reserves, where the goal of management is the maintainance of "natural" conditions. "Let-burn" policies, which allow fires to burn unless they threaten structures or developed areas, have been initiated in many National Parks and Monuments since the late 1960's (Kilgore 1976), and more recently in wilderness areas of the National Forests (Davis 1979). "Prescribed burns," purposefully set fires, have been used for recreating colonial conditions in certain areas of western National Parks since the late 1950's (Kilgore 1976; Van Wagtendonk 1975). The most famous of these projects have involved the giant sequoia groves *(Sequoiadendron giganteum)* in Sequoia-Kings Canyon National Parks, in which an attempt has been made to reduce the density of forest stands and increase reproduction by the giant sequoias (Kilgore 1973).

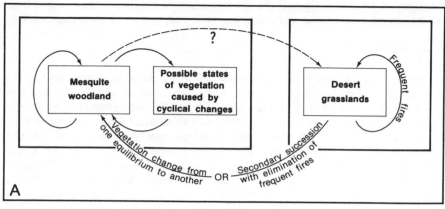

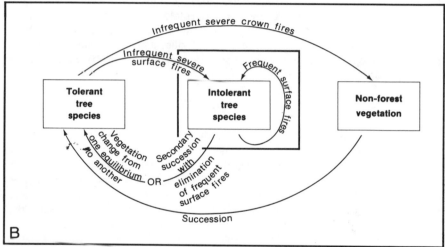

FIGURE 11   VEGETATION CHANGES ASSOCIATED WITH AL-
TERED FIRE REGIMES. Solid lines indicate different definitional bound-
aries for "vegetation" or "vegetation system." **A.** Desert grassland. **B.**
Forest.

Vegetation changes associated with the alteration of a fire regime may be
evaluated as either permanent (the changes represent the creation of a new equilibrium) or temporary (the changes represent fluctuations around an established equilibrium). These differing assessments are strongly influenced by how much vegetation
change one recognizes as part of the established equilbrium. For example, the expansion of mesquite *(Prosopis juliflora)* into the desert grassland is usually described as a
change not easily reversed because fires kill most young mesquite trees but few mature
individuals. Thus, the mesquite woodland which results from long-term fire suppression
may be described as a new equilibrium that has replaced the grassland. On the other
hand, this transformation of the plant cover may also be seen as but one possible
change within the equilbrium condition of the desert grassland (Figure 11a). Similarly,
in forested environments (Figure 11b) where increases in tree density have permitted
crown fires to burn where formerly surface fires were characteristic, a new disturbance

regime can be identified. Conversely, the less frequent but more severe crown fires could be viewed as a variation within a larger-scale portrayal of the vegetation system.

Livestock Grazing

The literature on relationships between vegetation and grazing is large, but most of it is prescriptive. Articles report the effects of given livestock use on certain plant species (White and Terry 1979; Painter and Detling 1981) or on vegetation types (Pitt and Heady 1979; Laycock and Conrad 1981), with either expressed or implied recommendations for management of livestock or range vegetation to achieve desired production goals. Less attention is paid to exploring how vegetation characteristics may reflect livestock use, in the past or present, although such studies do exist (Burcham 1957; Vale 1975a; Branson and Miller 1981). The economic importance of the range livestock industry explains this bias, and, in fact, the prescriptive approach reveals much about the ecology of grazing which then may facilitate interpretations of vegetation characteristics and patterns. However, a recent statement about the paucity of data concerning grazing effects on lands administered by the Bureau of Land Management is true more generally: "A disturbing fact is that little, if any, quantitative data on vegetation changes, toward improvement or deterioration, is available for public domain lands" (Branson and Miller 1981:9).

Ecology of Grazing

Grazing by domestic livestock influences wild vegetation directly through removal of foliage and indirectly by trampling of both plants and soils (Lewis 1969; Heady 1975; Stoddard et al. 1975). Grazing and browsing by native animals (Batzli and Pitelka 1970; Ross, Bray, and Marshall 1970; Snyder and Janke 1975) and feral species (Bratton 1975; Hanley and Brady 1977) also influence the character of vegetation. Domestic livestock tend to graze selectively on vegetation and, thus, do not duplicate the qualitative or quantitative features of more general foraging by a group of native ungulates (Myers 1972). Nonetheless, the strong competition between cattle or sheep and wild animals (Mackie 1978) suggests that duplication in feeding characteristics exists.

Carbohydrate storage cycle. By photosynthesis, plants transform the radiant energy of sunlight into latent energy of biomass, and store the created carbohydrates for subsequent respiration. During a dormant season, carbohydrate reserves are gradually reduced. When the growing season begins, a rapid but usually short reduction occurs, until the plant's activity allows a subsequent net accumulation of carbohydrates. Flowering and seed production typically require stored energy, but later in the growing season, carbohydrates again accumulate, to be drawn upon during the following dormant season.

Grazing may influence the amount of energy available to plants by reducing the photosynthetic area of the leaves. Thus, grazing can reduce the vigor or the reproduction of the affected plant if the grazing is heavy and occurs at a time when the plants are at a vulnerable stage in the carbohydrate cycle. The vulnerable times differ among species, but the result may be similar: a decrease in importance of grazed species in the vegetation.

Plant Morphology. The physiology of carbohydrate storage influences the response of plants to grazing, and variations in the physical morphology of plants are also

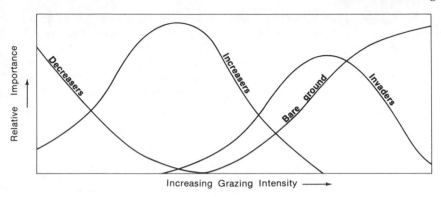

Increasing Grazing Intensity ⟶

FIGURE 12   GENERALIZED RESPONSE OF VEGETATION
TO A CONTINUUM OF GRAZING PRESSURE

important. For example, grasses with an abundance of vegetative stems and axillary buds resprout better when grazed than species with few vegetative stems or axillary buds. Such differential response to grazing has been invoked to explain the increase in sod-forming short grasses and decrease in bunch form grasses in parts of the western North American prairies since European settlement.

Root growth. Grazing influences growth of roots as well as leaves. Heavy grazing may, in fact, prevent root growth entirely by reducing leaf area and hence energy assimilation. Poor root development restricts carbohydrate storage capacity of plants and thus decreases their vigor.

Reproduction. Light grazing may actually increase vegetative reproduction of grasses by encouraging tillering. Heavy grazing reduces sexual reproduction by inhibiting the amount of carbohydrate stored which in turn reduces seed production. In addition, grazing when seed heads are immature typically decreases the seed crop, although grazing after seeds are fully mature may help disperse seeds and encourage their germination by trampling them into the soil.

Annual productivity. With increased tillering caused by grazing, sod-forming grasses may produce more biomass in a season with light livestock use than without grazing. Annual yields of bunch grasses, on the other hand, are typically higher with a single clipping at the end of the growing season than with multiple harvests.

Nutritive content of forage. New plant growth usually has a higher content of such nutrients as protein, potassium, and phosphorus. Thus, to the degree that grazing increases reproduction but does not restrict annual productivity, forage nutrients may be increased with livestock use.

Effects of Grazing

Range managers often characterize the effects of grazing in terms of vegetation responses on a continuum of grazing intensity (Figure 12). The initial effects of grazing are physiological set-backs to the most palatable species. If grazing intensity is increased, these species decrease in importance (such species are called "decreasers"), while other less palatable species increase ("increasers"). More grazing pressure results in a decline in the "increasers" and invasion by still less palatable plants ("invaders"), which are often woody or annual. If livestock pressure grows still greater, the total coverage of the vegetation often begins to decline, first among plants in the pre-grazing vegetation and then among invaders. The end point on the continuum is the total loss of plant coverage, or exposure of bare ground. This series or continuum is usually called "retrogression" of the vegetation. The classification of a particular plant species may vary depending upon the preferences of different grazing animals.

TABLE 2　CHARACTERIZATION OF PLANT SPECIES RESPONSE TO LIVE-STOCK GRAZING ON DESERT GRASSLAND IN SOUTHERN ARIZONA[a]

Plant Species	Percent Coverage After Different Grazing Histories[b]			Characterization[c]
	Heavily Grazed Before 1941	Less Heavily Grazed After 1941	Not Grazed After 1941	
GRASSES				
Arizona cottontop (Digitaria californica)	0	0.08	0.70	Decreaser
Sideoats grama (Bouteloua curtipendula)	0	0.01	0.53	Decreaser
Poverty threeawns (Aristida divaricata and A. hamulosa)	trace	0.32	0.37	Decreaser if species remains more important without grazing than with grazing; increaser if species declines without grazing
Rothrock grama (Boutelova rockrockii)	0.06	0.52	0.35	Increaser
Red threeawn (Aristida longiseta)	trace	0.07	0.34	Increaser if species remains without grazing; invader if species disappears without grazing
Other grasses	0.06	0.12	0.71	Collectively behaving as decreasers
TREES AND SHRUBS				
Velvet mesquite (Prosopis juliflora var. relutina)	0.76	5.32	3.80	Decreaser in the absence of hot fires
Burroweed (Haplopappus tenuisectus)	6.38	2.30	1.69	Invader, assuming further decline without grazing
Sticky snakeweed (Gutierrezia lucida)	0.84	1.10	trace	Invader
Wright buckwheat (Eriogonum wrightii)	0.02	0.16	2.48	Probably an increaser because species declined in frequency without grazing
Other shrubs	0.52	0.79	trace	Collectively behaving as increasers

[a]Adapted from Schmutz and Smith (1976) by permission of E. M. Schmutz and Society for Range Management. From "Successional Classification of Plants on a Desert Grassland Site in Arizona," E. M. Schmutz and D. A. Smith in *Journal of Range Management* 29 (1976): 476-479.
[b]Coverage for grasses is area of basal tufts; for woody plants it is area of crown
[c]See text for discussion of terms

Changes in vegetation composition. Grass species in various North American grasslands have been classified according to their responses to grazing (Table 2). Assuming that species responses are consistent from place to place, range managers often use composition of grasslands as an index of past or present grazing intensity. Implicit in this technique of range evaluation is the idea of a general "climax" or equilibrium vegetation from which the plant cover has deviated under the influence of livestock grazing. Vegetation other than grasslands has been similarly evaluated by comparing areas grazed or browsed with areas not used by livestock, or by identifying the trends in previously-grazed vegetation that has been given protection (Blydenstein *et al.* 1957; Marquiss and Land 1959; Schmutz 1967; Schmutz and Smith 1975).

Deterioration of woodlands. Change in vegetation composition caused by grazing does not always act to increase woody plants. In some western woodlands, livestock use has contributed to a lack of tree reproduction (Boldt *et al.* 1978; Crouch 1979; Shanfield 1981). In a parallel way, browsing by native ungulates has suppressed aspen regeneration and helped transform woodlands in the northern Rocky Mountains to coniferous forests or to non-forested plant covers (Krebill 1972: Gruell and Loope 1974). Schier (1975) has warned, however, that it is easy to blame livestock for woodland deterioration when, in fact, other factors may be more important.

Ecotone Movements. Compositional changes in range vegetation caused by grazing may take the form of an invasion by trees into vegetation where they were formerly excluded. In such situations, the pregrazing vegetation is envisioned as having been "closed" to invasion — available resources were fully utilized — and livestock use "opens" the vegetation to tree establishment (Robertson and Pearse 1945). In sagebrush-grass vegetation, Vale (1975b, 1977) argued that grazing was responsible for tree invasion, although Burkhardt and Tisdale (1976) invoked suppression of fires, and Blackburn and Tueller (1970) thought both factors were important. In mountain meadows of the West, grazing, particularly by sheep, is often cited as responsible for tree invasion (Runnell 1951; Vankat and Major 1978; Boche 1974; Vale 1981), but both fire suppression (DeBenedetti and Parsons 1979) and climatic fluctuation (Wood 1975) are also suggested. Interactions among these factors and their coincident timing often make isolation of a single cause of tree invasion difficult or inappropriate (Vale 1981).

Changes in fire characteristics. The invasion of desert grasslands by woody plants provides an example of the interplay between grazing use and fire characteristics. Some researchers have suggested that grazing alone has not been directly responsible for expansion of woody plants in that environment; rather, the grazing has reduced the fuels for fires, decreased the frequency of burning, and thereby allowed establishment of trees (Wright and Bailey 1980).

Changes in Equilibria

Livestock grazing is a factor which, like climate or fire, influences the composition and structure of vegetation. Grazing may be viewed as disturbance, moreover, thereby permitting the various interpretations of vegetation change caused by disturbance events (Figure 13). A plant cover may achieve a kind of stability under a given regime of grazing intensity and type, in which the composition and structure are maintained through time (Odum 1959; Beetle 1974). With complete elimination of livestock, the vegetation may change to reflect the altered conditions. That change may take one of two forms. First, the plant cover may "progress" away from the disturbed state caused by grazing and toward a regionally-identifiable stable condition. Such development

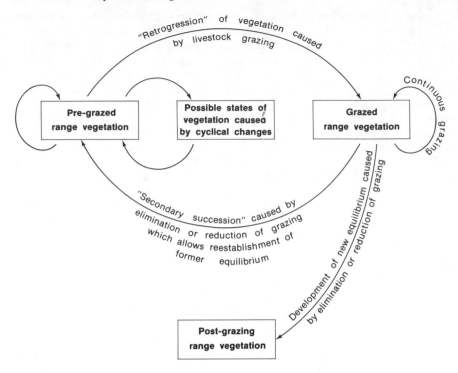

FIGURE 13   GENERALIZED VEGETATION CHANGES
ASSOCIATED WITH LIVESTOCK GRAZING

would retrace the retrogression continuum. This directional change conforms to classic concepts of succession and ends with a stable plant cover equivalent to a regional climax which may have existed in pre-grazing times.

The second form of change is toward an equilibrium that is not identical to that existing prior to livestock use. The plant cover of a particular area may develop toward a stable state that reflects the species present when the grazing was ended or reduced, the availability of potential colonizing species, and the environmental conditions of the site at the time of change in livestock use. For example, Anderson and Holte (1981) described the development of rangeland in southern Idaho after elimination of livestock and found that the composition of the vegetation varied from place to place, reflecting the array of plant species present when grazing was ended. No successional sequences could be identified, and many different compositions were apparently stable. Increased sagebrush density caused by grazing in sagebrush-grass vegetation is sometimes described as irreversible, even with an elimination of cattle or sheep, and thus represents a new stable plant cover (Pechanec *et al.* 1965). Forest establishment in mountain meadows would be another example.

Vegetation may also change with reduction in grazing intensity or changes in types of animals being grazed. Such changes may also be site specific, and, thus, the variety in stable configurations achieved will preclude the identification of a regional "climax."

Logging

Most of the literature on logging, like that on grazing, is prescriptive. Innumerable studies report the results of specific logging procedures on particular tree species, vegetation stands, or environmental characteristics, with recommendations for management (Oliver 1979; Alexander and Edminster 1980; Metzger 1980). Surprisingly few works evaluate past or contemporary logging as an ecological factor that has influenced, or is influencing, vegetation patterns and characteristics in regional environmental settings.

Several examples of research problems that evaluate logging as an ecological factor can be cited. Eastern white pine *(Pinus strobus)* and red pine *(Pinus resinosa)* were apparently more important in the Great Lakes region in presettlement times than today, although a quantitative description of their decline has not been made. Why these pines should not be reproducing well today, even on clearcuts or after fires, is not well documented. Ahlgren (1976) suggests that lack of seed sources and expansion of competing aspen are probably important. Both of these factors, in turn, seem related to early, heavy logging which eliminated pines from many areas and encouraged the spread of aspen. Ahlgren's suggestions may well have regional importance accounting for the impact of early logging on modern forests in the northern hardwood-white pine-hemlock region.

A second example involves the heavy demands for wood both for fuel and construction materials which characterized early mining regions in the American West. In areas without abundant forests, this need undoubtedly led to local deforestation. Need for fuel and agricultural timber resulted in elimination of juniper stands near Salt Lake City (Christensen and Brotherson 1979). A similar pattern undoubtedly was repeated throughout the intermountain west. Betancourt and Van Devender (1981) have even suggested that native Americans may have caused local deforestation in the semi-arid West.

In California, coast redwood *(Sequoia sempervirens)* has long been both logged and admired aesthetically for its great size (Simmons and Vale 1975). Demands for parks typically have been based on the allegation that timbercutting has decreased the abundance and distribution of the species. No one, however, has assessed the importance of more than a century of logging on coast redwood. Such an inquiry would be particularly useful in regions where good early documentation of redwood stands is available. These data could be compared with early logging, contemporary timber harvesting, and the character of the present vegetation.

Ecology of Logging

Timber cutting has effects which parallel other causes of disturbance in forests. Variations in logging intensity determine the volume of wood removed and the related degree to which the forest canopy is opened. These variations form a continuum but are usually classified into logging types as exemplified by the names of silvicultural systems used by modern foresters (Figure 14).

Silvicultural systems. Logging which removes single mature trees ("individual selection") or small patches of trees ("group selection") continuously maintains a forest with trees of all ages, and thus is called "uneven-aged" management. The disturbance created by logging is relatively modest, with reproduction concentration in the small openings created by the timber harvesting.

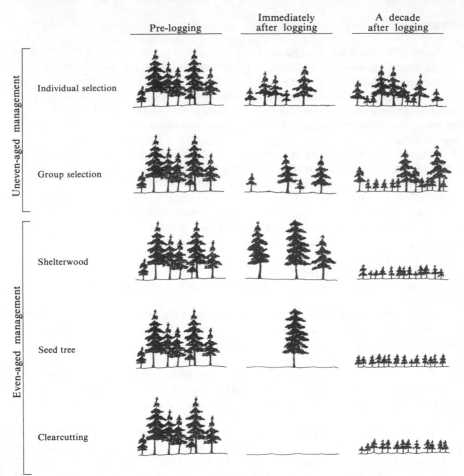

FIGURE 14   SILVICULTURAL SYSTEMS. Redrawn from Stephen H. Spurr, "Silviculture." Copyright © 1979 by Scientific American, Inc. All rights reserved.

Logging that removes most or all trees, regardless of size, over relatively large areas creates conditions for a regeneration of the entire forest. The new trees germinate about the same time, and thus the resulting stand and the type of management are called "even-aged". A shelterwood cutting removes all but a portion (e.g., 25%) of the canopy trees which remain as protective cover for the regeneration. Once the new trees are established, the shelterwood is also cut. Seed tree systems remove all trees except for a few scattered individuals that serve as seed sources for the new forest. Clearcutting is the most extreme even-aged system. All trees are cut, and a new stand develops from seed from unlogged areas or sprouts from cut trees. In each case, even-aged management is a catastrophic disturbance of the forest.

Early logging practices probably approximated the heavy cutting characteristics of modern even-aged management. Contemporary timber harvesting in the United States is much more typified by the even-aged systems than by selection systems (U.S. Senate 1971), although the adoption of a logging procedure in a particular situation depends upon many considerations.

Reasons for selecting silvicultural systems. Depending upon the degree of disturbance caused by logging, different species can be expected to regenerate in a forest. Forest managers who wish to perpetuate or increase tree species that are intolerant of shade or otherwise dependent upon disturbance usually advocate the heavy cutting systems of even-aged management. On the other hand, if species which reproduce best in undisturbed conditions are desired, selection cutting is considered better.

If a mature forest is naturally even-aged, such as one that became established after a widespread crown fire, foresters often argue that selection systems are inappropriate because mature trees left unharvested will decline in vigor and eventually die. Heavier cutting thus maximizes the productivity of the forest for wood products. Similarly, selection cutting is innately more applicable for uneven-aged stands, although other considerations may encourage heavier cutting systems in such forests.

Weather factors may favor certain harvesting systems. Shelterwood cutting rather than clearcutting is sometimes used in environments where young trees might be subject to damaging frost, such as inland forests of Douglas fir (U.S. Forest Service 1973). The potential blow-down of trees left standing after logging, on the other hand, might prompt a heavy cutting in order to minimize waste. Such a rationale has been used to justify clearcutting in forests of coast redwood (Boe 1975).

Still other factors influence the choice of a particular silvicultural procedure. For example, a forest stand that is heavily-infested with a pest, such as dwarf mistletoe in lodgepole pine, is often recommended for heavy cutting in order to reduce spread of the pest organisms and salvage as much of the forest as possible for wood products. Concern for wildlife, watershed functions, or aesthetics, moreover, may prompt adoption of silvicultural systems that would not be dictated by the desire for timber products alone.

The economic attractiveness of differing silvicultural systems influences the choice of a particular timber harvesting technique. Heavier cutting often produces higher economic returns than lighter cutting simply because of the larger volume of wood harvested. Some environmental groups charge that economic considerations dominate silvicultural decisions.

Three examples of silvicultural system effects. The mountainous Pacific Northwest supports vast areas of Douglas fir *(Pseudotsuga menziesii)*. These forests originate after crown fires produce mineral seedbeds and strong overhead light that the species requires for reproduction. Logging is usually clearcutting, often with post-logging burning of slash and planting of young trees. Seed tree systems offer little advantage because of the artificial planting, and the mature trees left standing are allegedly subject to windthrow. Shelterwood harvesting is practiced in interior locations where cold temperatures are potentially damaging to young trees. Individual selection systems produce too little disturbance to allow establishment of young Douglas fir, and instead encourage more shade-tolerant species such as western hemlock *(Tsuga heterophylla)*. Commercial clearcutting in Douglas fir is often described as a means of replicating the natural catastrophic burns which permitted the initial establishment of these forests prior to European contact. This ecological interpretation of clearcutting is supported by studies of forests in New England (Bormann and Likens 1979). Other observers, in contrast, have argued that group selection provides disturbance sufficient for Douglas fir reproduction, while also maintaining a continuous forest cover with

TABLE 3   EFFECTS OF TWO SELECTION CUTS ON A MIXED-COMPOSITION HARDWOOD FOREST IN WEST VIRGINIA

Tree Species	Percentage of Trees Within Size Class (inches)					
	1 to 5		5 to 11		Over 11	
	Before cutting	10 years after cutting	Before cutting	10 years after cutting	Before cutting	10 years after cutting
Most Tolerant Species						
Sugar maple	42.7	42.8	7.8	16.0	7.4	5.8
(Acer saccharum)						
American beech	33.8	34.2	6.9	9.8	15.4	11.3
(Fagus grandifolia)						
Red maple	10.3	9.8	7.7	10.7	2.9	4.0
(Acer rubrum)						
Less Tolerant Species						
Sweet birch	4.6	1.8	12.5	13.2	4.7	4.6
(Betula lenta)						
Red oak	2.6	1.5	14.1	10.0	15.6	24.2
(Quercus rubra)						
Black cherry	0.0	0.0	7.6	4.2	17.2	16.9
(Prunus serotina)						
Yellow poplar	0.0	0.0	5.3	3.8	12.5	14.4
(Liriodendron tulipifera)						

Source: McCauley and Timble 1975:5.

greater diversity in age of trees and understory characteristics (Twight 1973).

Individual selection harvesting in hardwood forests of West Virginia (Table 3) seems to accelerate the tendency of shade-tolerant trees to dominate stands (McCauley and Trimble 1975). The overstory prior to logging included shade-intolerant species, particularly red oak (Quercus rubra), probably established after earlier heavy logging and burning. The saplings, on the other hand, were mostly shade-tolerant trees, notably sugar maple and American beech. After two cuts over a ten-year-period, the species in the sapling layer were unchanged, but the composition of the intermediate-sized group shifted away from shade-intolerant trees to shade-tolerant species. Apparently, the progression toward a sugar maple-red maple-beech forest was not altered by the minor disturbance of an individual selection system. Reluctant to interpret such findings as an endorsement of clearcutting, Twight and Minckler (1972) suggested that group selection will create conditions suitable for establishment of shade-intolerant trees while not creating what they described as clearcutting's undesirable effects (large areas of homogeneous forests, or expanses of non-forest terrain after harvesting).

An interesting evaluation of the different effects of silvicultural systems on mixed coniferous forests is provided by McDonald (1976). Prior to logging, an experimental forest in California was dominated by three coniferous species: Douglas fir, ponderosa pine, and incense cedar (Calocedrus decurrens). The largest trees were mostly ponderosa pine (Table 4). Several broadleaf species were also important. Nine years after logging, the composition of the reproduction varied with the silvicultural system. Clearcuts were covered by dense stands of brush with ponderosa pine. Seed tree and

TABLE 4 SIZE STRUCTURE OF A CALIFORNIA MIXED CONIFER FOREST PRIOR TO LOGGING

Tree Species	Number of trees per acre by size classes (diameter in inches)			
	3.5-12.0	12.1-20.0	20.1-30.0+	Total
Douglas Fir (Pseudotsuga menziesii)	40	10	6	56
Ponderosa pine (Pinus ponderosa)	6	14	23	43
Sugar pine (Pinus lambertiana)	1	1	2	4
White fir (Abies concolor)	8	2	1	11
Incense cedar (Calocedrus decurrens)	59	5	1	65
Broadleaf trees	60	7	2	69

Source: McDonald 1976:4.

shelterwood cuttings supported the densest reproduction recorded for both ponderosa pine and broadleaf trees. Sugar pine (Pinus lambertiana), an intermediate-tolerant species that requires more mesic conditions than ponderosa pine, achieved its highest numbers after shelterwood cutting. Group selection resulted in fewer ponderosa pine and broadleaf species, although sugar pine remained relatively numerous. The shade-tolerant white fir (Abies concolor) seeded best after group-selection, and it was joined by Douglas fir after individual selection cutting.

All trees grew faster on clearcuts than in any other silvicultural system, but the relative growths were different (Table 5). Ponderosa pine grew faster than the other conifers in clearcuts, and white fir grew most slowly. Conversely, ponderosa pine grew most slowly in the individual selection forest, and white fir grew fastest. Both the composition and the growth of the young trees is thus dependent upon the degree of disturbance. These reactions also suggest that the pre-logging forest was probably open and sunny, judging from its domination by ponderosa pine.

TABLE 5 SEEDLING DENSITY IN RELATION TO SILVICULTURAL SYSTEM, NINE YEARS AFTER LOGGING, CALIFORNIA MIXED CONIFER FOREST

Tree Species	Silvicultural System				
	Individual Selection	Group Selection	Shelterwood	Seed tree	Clearcut
Ponderosa pine	860	1500	3620	2100	1115
Sugar pine	111	185	240	75	51
Douglas fir	308	134	80	174	157
White fir	400	565	192	66	166
Incense cedar	44	16	470	67	0
Broadleaf trees	1330	807	2225	2937	746
Broadleaf shrubs	0	0	0	0	6523

Source: McDonald 1976:6.

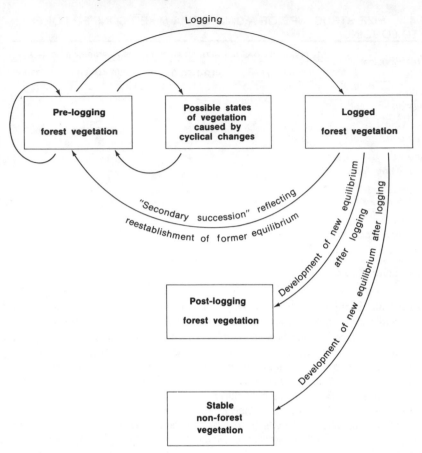

FIGURE 15   GENERALIZED VEGETATION CHANGES
ASSOCIATED WITH LOGGING

Logging as a process is similar to other causes of major disturbance in forests (Bormann and Likens 1979). Depending upon the degree of disturbance, species that respond favorably to the conditions created by timber harvesting will increase. The pre-logging forest may be reestablished, with the vegetation progressing in a way which resembles classic succession (Bormann and Likens 1979; Gashwiler 1970; U.S. Forest Service 1971). On the other hand, post-logging development of the plant cover may result in new equilibria being established. Poor regeneration of forests on climatically or edaphically marginal sites produces essentially stable plant covers of herbaceous or brushy vegetation (Bormann and Likens 1979; Minore 1978; Antos and Shearer 1980). Even where trees do regenerate, the composition of the post-logging forest may be different from that existing prior to timber cutting because of chance availability of seeds (Antos and Shearer 1980), or elimination of seed sources by the logging itself (Ahlgren 1976). A diagram of vegetation change associated with logging is thus similar to that for livestock grazing, except that the processes identified involve removal of timber rather than forage (Figure 15).

Trampling

A growing literature evaluates the effects of recreational trampling, mostly by humans but also by livestock, on vegetation. These studies typically adopt one of three approaches: comparing sites presumed to be heavily trampled with sites which are not, evaluating vegetation that is newly exposed to or recently protected from trampling pressure, or documenting the impacts of experimentally induced trampling. The vegetation in each of these approaches may be influenced not only by the direct effect of trampling, mechanical damage to plants, but also by indirect effects of trampling such as increased soil compaction or reduced competition resulting from the elimination of species sensitive to treading. Moreover, other environmental changes not caused by, but associated with, trampling may alter the plant cover. These include changes in microclimate produced by clearing a trail corridor in a forest, increases in available nitrogen from the excretions of recreational livestock, and higher grazing pressures beside trails used by horses or mules. Thus, assessments of trampling often include many factors in addition to the direct mechanical effects of treading.

A diagram helps to organize the factors which influence trailside vegetation, where the complication of multiple factors is probably greatest (Figure 16). The trail center is normally bare of plant growth due to intense trampling pressure (Bates 1935; Bayfield 1973; Liddle 1975). Immediately beside the trail center, certain plant species are less important than away from the trail as a result of vulnerability to the mechanical damage

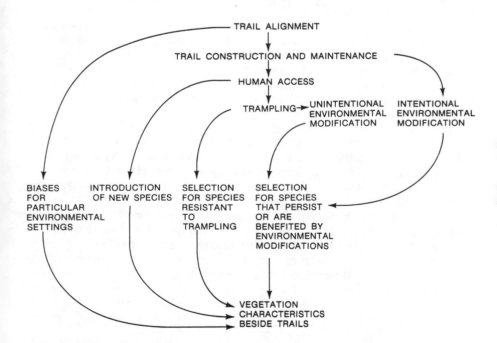

FIGURE 16   FACTORS INFLUENCING CHARACTERISTICS OF TRAILSIDE VEGETATION (Modified from Liddle 1975 and Cole 1978; reproduced by permission of Applied Science Publishers Limited and Blackwell Scientific Publications Limited)

of trampling; woody plants are disproportionately represented in this group (Dale and Weaver 1974). Other species with structural features (short, flattened basal leaves; hemicryptophyte form) or behavioral characteristics (annual, early-maturing) which render them less vulnerable to trampling may increase (Bates 1935; Dale and Weaver 1974; Liddle and Greig-Smith 1975; Holmes 1976). Species may also respond differently to the unintentional changes associated with a trail, including, for example, an increase in invasive plants but a decrease in species that require undisturbed substrates (Willard and Marr 1971; Dale and Weaver 1974). In sum, the plant cover at the trail edge should normally grade into vegetation of the untrampled environment with increasing distance away from the trail, producing patterns resembling those of plants beside and away from highways (Frenkel 1974).

Complicating this picture of unintentional modification of the plant cover are activities by which people intentionally alter vegetation and soils along trailsides. Examples of these factors, which have been largely ignored in the trampling literature, include conscious selection for certain habitats in choosing an alignment for a trail (presence of soil and lack of rocks in western forests, or presence of grass rather than shrubs in heather vegetation — Boorman and Fuller 1977), construction of a trail with unobstructed clearance and good grades, maintenance of particular ground surfaces on a path (mineral soil in trails of the mountainous West, or coverings of wood chips in some heavily-used areas), or the removal of rocks and woody plants from trail corridors by trail crews. Sampling vegetation along trailsides thus reveals a plant cover that reflects more forces than those associated with trampling or associated unintentional modifications of the environment.

In contrast to the analysis of trailside vegetation, experimental trampling usually excludes many alterations of the environment found beside trails, and thus gives an incomplete view of the forces at work in recreation areas. However, the experimental approach allows the isolation of the trampling factor and permitted Weaver and Dale (1978), for example, to conclude that trails on slopes are wider than those on the level, horses produce wider trails than hikers, and trails in forests tend to have a higher percentage of bare ground than those in grasslands.

Roadside campgrounds and back-country campsites are, like trails, subject to trampling (Gibbens and Heady 1964; LaPage 1967; Magill 1970; Foin et al. 1977; Cole 1981, 1982). The effects are similar to those produced along trails, although the related factors may be different. For example, in campgrounds the effects associated with recreational livestock are normally low or absent, but the cutting of trees and shrubs, and the collecting of wood for fires, may be high.

Environments vary in reaction and sensitivity to trampling pressure. Alpine tundra is damaged by even low levels of human treading (Willard and Marr 1970, 1971). Sedge meadows, in contrast, may be relatively resistant to heavy foot traffic (Foin et al. 1977). Dune vegetation subject to moderate trampling may experience increased availability of soil moisture due to soil compaction (Liddle and Greig-Smith 1975). Cole (1979) has argued that forest vegetation is generally more sensitive than that of meadows, thereby calling into question the policy of moving trails out of western mountain meadows. Moreover, Cole (1978) has also suggested a technique with which to estimate the vulnerability of plant covers to trampling.

With continued constant levels of trampling, the vegetation may develop a new equilibrium that reflects trampling intensity in the same way that livestock grazing may maintain a certain vegetation cover on rangelands (LaPage 1967). Elimination or reduction of trampling should normally allow vegetation to change toward some other equilibrium. This change may resemble secondary succession, but it may also create a plant cover not identical to pre-trampling conditions. A badly eroded trail, for example (Helgath 1975; Coleman 1977), may not have the soil conditions existing prior to the trail, and thus the vegetation that develops may be different. Species available for colonization may likewise influence post-trampling vegetation development. The same vegetation changes resulting from other disturbances also occur with trampling.

Off-Road Vehicles

Several types of off-road vehicles — snowmobiles, trail motorcycles, and 'dune buggies' — are important modifiers of vegetation in certain recreational areas. When their use is restricted to tracks or trails, off-road vehicles have an impact that is narrow and linear, much like the pattern of trampling effects along foot or horse trails. In some locales, on the other hand, particularly in southeastern California deserts, the use of off-road vehicles is less constrained and thus impacts are more widespread.

More than one-half of U.S. registered snowmobiles are in the Great Lake States of Minnesota, Michigan, and Wisconsin (Brander 1974). Little research has been published evaluating the effects of snowmobiling on vegetation, although parallels with the impacts of trampling are obvious. The vehicles break, reduce, or eliminate woody plants which protrude from the snow surface (Neumann and Merriam 1972). Where the snow cover is thin, damage to the plant cover increases, with decreasing coverage and increasing exposure of bare ground. Plants least susceptible to damage are ground-hugging, herbaceous, and sod-forming (Greller *et al.* 1974). Compaction by snowmobiles reduces snow's insulation capacity, with increasing potential for cold temperature damage to organisms on or in the ground (Neumann and Merriam 1972; Brander 1974).

Off-road vehicles which travel on the ground surface readily destroy vegetation (Davidson and Fox 1974; Luckenbach 1975; Bury *et al.* 1977; Hosier and Eaton 1980). On trails, Weaver and Dale (1978) found that after the same number of passes by small, conservatively driven motorcycles and conservatively ridden horses, the horses had a greater overall impact on vegetation and soils. In desert environments and on dunes, however, off-road vehicles have undeniably caused wide-spread vegetation destruction (Luckenbach 1975; Hosier and Eaton 1980). In arid regions, off-road vehicles reduce the plant cover not only directly by mechanically breaking woody plants, but also by compacting the soil (Davidson and Fox 1974; Eckert *et al.* 1979), which in turn reduces water infiltration properties and the successful germination of annual plants (Luckenbach 1975; Davidson and Fox 1974; Iverson *et al.* 1981) Accelerated erosion on the compacted soil (Eckert *et al.* 1979) may also influence the plant cover.

Most concern has been expressed over the impacts of off-road vehicles in sensitive environments that are affected with even minor use and are slow to recover — arctic tundra (Rickard *et al.* 1974), alpine tundra (Greller *et al.* 1974), coastal dunes (Hosier and Eaton 1980; McAtee and Drawe 1981), and deserts (Stebbins and Cohen 1976). Also, much concern has focused on the effects of off-road vehicles on wildlife, either directly through physical damage or harassment, or indirectly through altered plant covers (Bury *et al.* 1977; Busack and Bury 1974; Eckstein *et al.* 1979; Neumann and Merriam 1972). For example, the numbers and diversity of small mammals and reptiles was found to be inversely related to off-road vehicle use in southeastern California (Bury *et al.* 1977).

The alteration of vegetation equilibria caused by off-road vehicles may be likened to the impacts of trampling or livestock grazing. Light or moderate use may initiate vegetation change and, then, maintain an equilibrium reflecting the passage of the vehicles. Heavy use may create exposures of bare ground that, too, may be perpetuated with continued intensive impact. Recovery to pre-disturbance conditions depends upon the degree of environmental alteration and the availability of species to recolonize affected areas. Because soils are typically changed, however, the reestablishment of vegetation identical to that existing prior to off-road vehicle use seems unlikely in most situations.

Air Pollution

Studies of the impacts of various types of air pollution on plant species and vegetation types are reported in a large literature, including many review articles and books (Muller and McBride 1975; Smith 1974; Mudd and Kozlowski 1975; W. H. Smith 1980; Treshow 1980). Most attention has focused on two pollution compounds, sulfur dioxide and ozone, although other substances (fluorides, heavy metals, limestone dust) have also been evaluated (Amundson and Weinstein 1980; Jordan 1975; Brandt and Rhoades 1973). Research has involved both laboratory experiments and field studies, but recently Legge (1980), after noting that most work has dealt with high concentrations of pollutants, commented that "very little emphasis has been placed upon integrated research programs concerning the impact of chronic, long-term, low-concentration air pollution stress."

Smith (1976) categorizes the effects of air pollution upon vegetation into three types, reflecting an increasing concentration of pollutants. With low pollution loads, plants act as sinks for the elements and compounds, with either benign or stimulating effects upon the vegetation. However, even low levels of heavy metals may have deleterious effects upon animals that eat contaminated plants. Intermediate pollution levels may interfere with plant physiology, resulting in reduced growth, reduced reproduction, and increased disease or insect attack. At still higher pollution loads, sensitive species may suffer increased mortality, and vegetation composition and structure may change as a result.

The most dramatic impacts of air pollution on vegetation are caused by high concentrations of atmospheric sulfur dioxide. Over sixty years ago, Hedgcock (1914) documented the deforestation associated with smelters in the Copper Basin of Tennessee. Since then other studies have reported similar severe effects (Gorham and Gordon 1960; Schofield et al. 1970; Linzon 1971; Wood and Nash 1976). A particularly good evaluation of sulfur dioxide impacts was provided by Gordon and Gorham (1963) at an iron-sintering plant at Wawa, Ontario, where the concentration of pollution decreased along a spatial gradient and thereby permitted identification of effects at different pollution loads (Table 6). The patterns are similar to those reported elsewhere: the highest levels of sulfur dioxide cause almost complete loss of vegetation cover; sensitivity of different portions of the vegetation to pollution is directly related to height or size of plants; some species increase in abundance at certain pollution loads because of the reduction of competing species; and plants differ in their ability to withstand pollution. The high sensitivity to pollutants of eastern white pine is often reported, whether caused by sulfur dioxide or ozone (Hayes and Skelly 1977; Skelly et al. 1979).

TABLE 6 CATEGORIES OF POLLUTION DAMAGE TO VEGETATION DOWNWIND FROM IRON SMELTER AT WAWA, ONTARIO

Damage Categories	Miles from Smelter	Trees	Characteristic Understory	Ground Vegetation	Erosion
Very severe	0-5	None	Almost destroyed; some *Sambucus pubens* alive, but damaged	Mostly destroyed; *Polygonum cilinode* present, but damaged	Severe; mainly bare rock in exposed locations
Severe	5-12	None	Much destruction; *Pyrus decora* nearly all dead; *Sambucus pubens* with dead tips	Ground vegetation dominant; *Polygonum cilinode* abundant; no tree seedlings, except a few *Betula papyrifera*	Apparent
Considerable	12-16	Canopy open; Some *Betula papyrifera* and *Picea glauca* alive	Tall shrubs dominant; most species with dead tips; some *S. pubens* present	Some damage; tree seedlings present	Not apparent
Moderate	16-24	Many conifers with dead tips; reduced crowns on hardwoods	Little damage; *Sambucus pubens* rare or absent	Normal woodland flora; *Polygonum cilinode* absent	Not apparant
Not obvious	More than 24	Overstory dominant; canopy closed; *Pinus strobus* with brown needles	Normal woodland flora	Normal woodland flora	Not apparent

Source: Gordon and Gorham 1963:1065. Reproduced by permission of the National Research Council of Canada from the *Canadian Journal of Botany,* Volume 41, pp. 1063-1078, 1963.

In terrestrial ecosystems, the most serious effect of "acid rain," pollution by relatively dilute concentrations of sulphur and nitrogen compounds, "may be the increased rate of leaching of major elements and trace metals from forest soils and vegetation" (Overrein 1980). More generally, acid precipitation seems to disrupt a variety of biochemical processes (Likens and Bormann 1974), perhaps contributing to declines in forest productivity (Cogbill 1976). Most recent concern has been expressed over the effects of acidic compounds upon aquatic ecosystems, particularly declines in fish populations (Gorham 1976). Large areas of crystalline bedrock, such as in Scandanavia, eastern Canada, northeastern United States, or parts of the mountainous West, which generally have low levels of carbonate compounds (which buffer acids and maintain neutral to basic pH levels), are particularly vulnerable to change by "acid rain" (Lewis and Grant 1980).

The effect of ozone on vegetation has been most thoroughly investigated in the mountains of southern California, particularly those surrounding the Los Angeles basin (Skelly 1980). Ponderosa pine is especially sensitive, with actual mortality of weakened trees usually caused by bark beetle attacks. That species may be replaced by more resistant trees in the future coniferous forests of the San Bernardino Mountains, and more resistant trees thus increase in importance (Cobb and Start 1970). Westmann (1979), moreover, has shown that the coastal sagebrush (Artemisia californica) scrub of southern California experiences reduced coverage with high oxidant concentrations.

In a thoughtful paper, Woodwell (1970) suggested that air pollution is equivalent in many ways to other environmental stresses, whether related to human activities (herbicide treatments, radiation) or to natural forces (high wind, low water, low energy). He indicated that in each of these stressful situations, a reduction in the size of dominant life forms occurs, probably because a plant of smaller stature can more easily maintain a favorable respiration/photosynthesis ratio. A decrease in productivity and in diversity also characterizes such environments.

New equilibria may result after vegetation has adjusted to given levels of pollution. With reduction in pollution loads, the characteristics of vegetation development ("recovery") may be complicated by the mix of species available for recolonization and by possible changes in soil conditions.

Construction Activity

Construction projects — laying underground pipelines; erecting towers for utility lines; removing, piling, and replacing overburden in strip mines; bulldozing road cuts — require removal of the plant cover and influence vegetation by destruction, creating areas of disturbed ground for recolonization. Vegetation development after such activities allows for the study of progressive vegetation change, whether or not it is classic "secondary succession."

Vegetation that occurs on sites disturbed by construction activity are, like those following most major disturbance events, disproportionately dominated by pioneer species, which are benefited by disturbance. In warm deserts, short-lived shrubs or perennial herbs occupy road edges, abandoned roadways, and sites disturbed by pipeline or power line construction. These shrubs persist in the "natural" desert in disturbed locales such as washes (Vasek et al. 1975a, 1975b; Jaynes and Harper 1978; Webb and Wilshire 1980; Lathrop and Archbold 1980a, 1980b). In Alaskan tundra, Chapin and Chapin (1980) reported that two sedges created a nearly complete cover on a bull-dozed plot after only five years, having spread from naturally disturbed ground produced by frost boils. On fifty-year-old alpine tundra road cuts, in contrast, pioneer bunch grasses created only one-half the coverage produced by "cushion" plants in undisturbed vegetation. These bunch grasses maintain themselves in the tundra environment, in part, by colonizing areas disturbed by ground-burrowing rodents (Greller 1974). Construction activity in forested environments also results in increases in herbaceous, shrub, or tree pioneers (Frenkel 1974; Ludwig et al. 1977; Anderson et al. 1977; Kruse et al. 1979; Johnson et al. 1979), those which would be associated with widespread or local disturbed sites produced by windthrow, fire, or landslides. In general, then, construction activities may be viewed as disturbance phenomena that encourage pioneer species in a manner similar to "natural" causes of disturbance.

Strip-mined locales and mine spoils have a type of construction-influenced vegetation that deserves special mention. Most of the exploding volume of literature dealing with mining and vegetation evaluates programs in which environments and plants are manipulated in order to establish a vegetative cover, rather than documenting vegetation development without direct human encouragement (Czapowskyi 1976). Nonetheless, Karr (1968) argued that over sixty years, revegetation from bare ground to shrub vegetation to forest occurred on strip-mined land in Illinois, apparently without human programs to promote a plant cover. In contrast, in the drier western states recolonization by vegetation is much slower (Wagner *et al.* 1978). Low levels of nitrogen or other essential elements and/or high concentrations of heavy metals act to retard natural revegetation even in relatively humid climates (Alvarez *et al.* 1974), and areas with both unfavorable climates and soils may have few if any vascular plants growing in mine spoils (Brown and Farmer 1976).

To the degree that post-construction vegetation development returns the plant cover to conditions which existed prior to disturbance, the old equilibrium will be attained. New equilibria may result from construction, however, if disturbance is maintained, such as on an unstable road cut. Also, vegetation development may progress toward new stable conditions of plant cover. These new equilibria may reflect changes in substrates, such as those associated with toxic materials in mined areas, or with compacted soils on abandoned roadways (Webb and Wilshire 1980). They may also develop as a result of the "pushing" of vegetation over an environmental "threshold," as in logged forests that fail to regenerate because of severe climates (Webber and Ives 1978). More simply, new stable vegetation forms may reflect non-successional but directional vegetation change. For example, Vasek *et al.* (1975b) suggest that both the season of disturbance and the availability of seeds influence vegetation development on power line sites. Thus, the plant covers which formed after construction activity varied from site to site, suggesting that a consistent and predictable successional sequence could not be identified.

Altered Biota

Alien plants introduced either accidentally or intentionally may naturalize sufficiently to appear to be a natural element in the plant cover. Annual grasses like wild oats *(Avena fatua* and *A. barbata)* have completely displaced native perennial bunchgrasses such as nodding needlegrass *(Stipa cernua)* (Bucham 1957, 1970). Salt cedar *(Tamarix pendantra* and *T. gallica)* has spread over flood paints in the Southwest (Robinson 1965); cheatgrass and other annual grasses have expanded across the northern intermountain West (Stewart and Hull 1949; Young *et al.* 1972; Young and Evans 1973). Buckthorn *(Rhamnus cathartica)* and bushy honeysuckle *(Lonicera tartarica)* dominate the understory of oak forests in parts of the upper Midwest (Jordan 1980). The climbing vine kudzu *(Pueraria lobata)* is an aggressive invader which has engulfed areas of the Southeast, and a multitude of non-native plants adorn roadsides and other disturbed locales throughout the continent. Such invasions and expansions of alien plants have received considerable attention (Elton 1958; Harris 1966; Jarvis 1979), with the role of human disturbance as a prerequisite an important theme (Baker 1972, 1974; Holzner 1978).

Introduced animals, particularly through foraging behavior, may alter plant covers. Wild boars in Appalachia (Bratton 1975; Singer 1981), goats on various islands (Spatz and Mueller-Dombois 1973; Coblentz 1977, 1978), deer in New Zealand (Veblen and Stewart 1980), and wild horses and burros in the arid and semi-arid West (McKnight 1958; Woodward and Ohmart 1976; Hanley and Brady 1977) all have had dramatic effects on the characteristics and patterns of vegetation.

Even the alteration of native animal populations by people is sometimes responsible for restricting the distribution of certain plant species or altering the structure of vegetation. Browsing by whitetail deer *(Odocioleus virginianus)* in the Great Lake forests, for example, may reduce the importance of eastern hemlock and increase the relative importance of sugar maple. With large numbers of deer encouraged by habitat modification and restrictive hunting regulations, the balance may be shifted away from hemlock and toward maple (Anderson and Loucks 1979). More generally, species palatability may make the "unnaturally" high deer populations a critical factor in forest composition (Ross et al. 1970; Bratton et al. 1980). In addition to influencing species mix, large numbers of ungulates may also reduce vegetation stature (England and DeVos 1969; Mueller-Dombois 1972; Snyder and Janke 1976; Marquis and Grisez 1978; Basile 1979).

Introduced diseases or pest insects have altered the composition of eastern forests in particular. Chestnut blight, Dutch elm disease (Karnosky 1979), and lethal yellowing palm disease (Fisher 1975) have each left a mark on the composition and appearance of tree-covered areas. Some of these introductions, such as the gypsy moth *(Porthetria dispar),* will have impacts over vast areas in years to come (Marshall 1981).

In the case of any alien introduction, or the alteration of native animal populations, the vegetation adjusts toward a new equilibrium. Ecologically, humans as dispersal agents of "non-natives" is neither new nor unique. But the pace of introductions by people, as with extinctions, may be much faster than the long-term natural norm.

Abandonment of Agricultural Land

Throughout most of the eastern states, large areas of forest cleared for agricultural purposes in the 18th, 19th, and early 20th centuries have been abandoned, and forests have reclaimed the former fields. The sequence of clearing, agricultural use, and invasion by trees has produced forests and woodlots that typically are different from those of precolonial times (Spurr 1979). In contrast, some Appalachian forests which have not been cleared and then reforested have compositions that are similar to those of 200 years ago (Lutz 1930; Spurr and Barnes 1980).

The theme of forests growing on abandoned farmland has been developed at two scales. The first scale deals with large regions and focuses on the changing area in forest. Clawson (1979), for example, compiled figures for the acreage in the United States in forest and farmland since 1800. The total area in forest declined until about 1920, then remained steady until after World War II, when it increased. A slight decrease has again occurred since 1970. This trend mirrors a complementary pattern in the area in farmland. The increase in forested acreage over the last several decades has not been paralleled by comparable increases in commercial forest, indicating that the abandoned farmland now wooded is not being intensively managed for forest products. The same trend of increases in forested area has been identified in "Megalopolis" by Gottman (1961), and by the U.S. Forest Service for individual states (Bones 1978; Powell and Kingsley 1980).

Delcourt and Harris (1980) have examined the role of forests in the southeastern states in the carbon cycle. They suggested that the area in non-forest land cover has increased, and the area in forest has decreased, since European settlement. Commercial forest land has accounted for a growing proportion of the total forested area, however, and commercial holdings have supported increasingly high rates of net annual growth. Thus, in spite of the smallest acreage in forests since 1750, the vegetation of the southeast is now accumulating carbon faster than it is releasing carbon to the atmosphere, reversing a trend toward ever lower amounts of stored carbon between 1750 and 1960.

The second scale at which abandoned farmland has been treated is local, focusing upon vegetation dynamics. The study of "old-field succession," secondary succession on abandoned fields, has a long history in plant ecology (Daubenmire 1968; Colinvaux 1973). Some studies have stressed the similarities between vegetation development on different sites, and have, thereby, identified characteristics of secondary succession. Others have emphasized the variable features of species replacing one another through time, due to such factors as differences in seed availability (Moss 1976), or conditions necessary for germination and survival (Ashby and Weaver 1970).

A particularly intriguing idea at this local scale is that farmers in parts of southern Appalachia practice a rotation in land use somewhat similar to slash-and-burn agriculturalists in the tropics (Hart 1977). Land cleared from forest for small fields is eventually abandoned in favor of newly cleared land. The old fields revert to woodland. Hart suggests that this ecological adaptation to soil depletion succeeds because of the modest economic goals of farmers who engage in the practice.

Purposeful Manipulation

The human activities discussed so far result in vegetation change, but the condition of the plant cover *per se* is rarely the purpose of any of the activities. Wild vegetation provides material products with logging or grazing, or the landscape is devoted to a particular use such as recreation or mining. In each case the plant cover may be changed, but the altered state is not the goal of the activity. Other human purposes, however, are served with conscious manipulation of wild vegetation in which the changed condition of the plant cover is the desired end. Even uncontrolled, "wild" vegetation reflects human handiwork.

Activities related to the practices already discussed include burning to reduce fuels for subsequent fires, often prescribed for forests and rangeland (Biswell 1963; Green 1981), and the reduction or elimination of woody vegetation in "fuel breaks," common in the brushfields of the mountains of southern California (Green 1977). The use of fire, machinery, or chemicals to reduce unpalatable plants and increase palatable grasses is a frequent practice on wild rangelands (Vale 1974; Roby and Green 1976; Kruse *et al.* 1979, Neuenschwander 1980).

Planting to reestablish a particular plant cover is probably the most common type of purposeful manipulation. Grasses are sometimes planted on burned brushfields or forests to reduce post-fire erosion; trees may be planted on clearcuts, such as Douglas fir stands (U.S. Forest Service 1973), or on burns (Bock *et al.* 1978) to accelerate reforestation. Crested wheatgrass *(Agropyron desertorum)* and other grasses are often established on grazing lands in the northern intermountain West (Vale 1974); and grasses are sometimes planted in heavily trampled recreation areas (Herrington and Beardsley 1970). The establishment of a vegetation cover on mine spoils or strip-mined areas is a particularly important example of planting in the wild landscape. Typically, post-mining revegetation requires suitable seedbeds, adaptable species, and irrigation or fertilizers to encourage survival. An "artificial" plant cover apparently will often persist on even environmentally difficult sites where "natural" revegetation is poor (Bay 1976; Day and Ludeke 1980; Brown and Johnston 1978a, 1978b; Monsen and Plummer 1978; Farmer *et al.* 1974).

Tree planting may be done for economic reasons, as with plantations of commercially valuable pines (Whitesell 1974; P.C. Smith 1980) or with windbreaks. In addition, aesthetic concerns may prompt people to establish trees, shrubs, or certain herbaceous plants in wild landscapes. The plantings may eventually seem an entirely "natural" part of the scene. Cottonwoods *(Populus fremontii)* have become an expected part of streamside vegetation in western Nevada even though they were apparently uncommon in precolonial times. Certain trees such as fan palms *(Washingtonia filifera)* around springs in the deserts of southern California likewise seem natural but in fact have been planted. Extensive stands of introduced *Eucalyptus* are so much a part of lowland California that they are an integral part of the image of the state. In another generation, groves of planted pines in southwestern Wisconsin may become indistinguishable from the naturally-occurring stands. Such naturalization of human-introduced species is most complete when the alien plant reproduces and spreads from the site where it was initially planted.

The logical extension of the activity of planting from wild landscapes to settled areas brings the discussion to urban forests and suburban gardens. Beyond the intended scale of this book, the theme of vegetation in cities and towns is little researched but extremely promising (Schmid 1975, 1979; McBride and Jacobs 1976; Rowntree and Wolfe 1980).

The goal of purposeful manipulation of vegetation is the establishment of a new equilibrium condition in the plant cover. Particularly important in this context is work which explores the problem of maintaining low-growing vegetation beneath power lines (Niering and Goodwin 1974; Egler and Foote, 1975). Short shrubs and herbaceous plants will resist tree establishment by utilizing resources needed by trees. Only in locales where the cover of low-growing plants is sparse will trees germinate and survive. Therefore, the desired low-stature vegetation is most easily maintained by eliminating single woody plants which are deemed troublesome, rather than by blanket application of herbicides or mechanical removal procedures which only act to "open" the vegetation to invasion by trees. In a theoretical sense, the concept of a relatively stable shrub cover, "closed" to tree invasion, is a refutation of classic views of succession, and instead emphasizes that whatever vegetation happens to develop on a site, as determined by a variety of factors, will remain until disturbance "opens" it to invasion by other plants. It is an endorsement of the view that vegetation equilibria are highly variable through space and time, rather than organized by cyclical changes.

Generalizations

no nearby powerlines

Several generalizations about human impacts on wild vegetation can be derived from the reviews of specific topics:

(1) Most simply, a human activity must be considered an environmental factor which, like other such factors, influences the dynamics of plant species. Some species may not be favored by the altered

conditions, and thus decrease in importance in the plant cover; others, however, may increase.

(2) Human activities usually disturb vegetation, with a resulting increase in plant species which are favored by disturbance. A major exception is the short-term effects of fire suppression, which is an impact of reduced disturbance. On the other hand, a long-term result of reducing fire frequency is often an increase in the magnitude of burns and thus in the intensity of disturbance.

(3) The disturbances caused by human activities replicate, in varying degrees, disturbances which result from "natural" processes.

(4) Intense disturbance may eliminate, at least temporarily, the plant cover.

(5) If the disturbance is continuing, the vegetation may adjust and then persist through time, reaching a new equilibrium.

(6) If the disturbance is not continuing, and if it is extensive, the human activity may initiate subsequent directional change in the vegetation in which the plant cover returns to a pre-disturbance equilibrium condition. This sort of directional change is usually called "secondary succession."

(7) Vegetation development after disturbance may also tend toward an equilibrium different from that existing prior to the human activity. This sort of change is analogous to the short-term directional, but non-cyclical, changes discussed in the first chapter.

(8) Human activities often result in vegetation with reduced stature. Diversity and productivity also may be changed, but the direction and degree of change are variable.

(9) Ecotones between forest and non-forest vegetation often are made unstable by human activities, with trees advancing into the non-forest vegetation.

Human impacts on vegetation, just as natural disturbances, are not limited to the vegetation alone. Vegetation is one set of elements in complex environmental systems. As vegetation is changed, so too can changes in other environmental processes be anticipated.

4

Effects of Altered Vegetation on Environmental Systems

Vegetation altered by human activities may influence other aspects of environmental systems, including wild animals, erosion, hydrologic characteristics, and soil fertility. Most vegetation changes described in Chapter 3 are similar in acting to disturb the plant cover. Discussions in this chapter will stress one major human disturbance activity, logging, for which the literature is particularly rich. Effects of other causes of vegetation change can then be compared to the impacts of timber cutting.

Effects on Wild Animals

Vegetation is the most important feature of the environment that influences the distribution and abundance of wild animals (Dasmann 1964; Leopold 1966). The altered vegetation resulting from human activities will necessarily have an impact on the wildlife associated with it.

Leopold (1966) provided a particularly good overview of the relationship between wild animals and human activities. He stressed that animal species vary in their habitat requirements, as well as in their ability to adapt to changed conditions. Indeed, some are favored by the environments created by people. Animals which prefer disturbed vegetation generally have fared better with European occupancy of North America than animals which depend upon undisturbed plant covers. Animal populations thus reflect the general influence of people as disturbance agents. Moreover, implicit in Leopold's discussion is the observation that species which benefit from human alteration include animals which are usually considered desirable, white-tailed deer *(Odocoileus virginianus),* moose *(Alces americana),* and mallards *(Anas platyrhynchos),* as well as species which are typically disliked, cockroaches *(Periplaneta americana)* or Norway rats *(Rattus norvegicus).* The image of people as solely destructive is an oversimplification.

Effects of Logging on Wild Animals

Large mammals. Several species of North American ungulates usually benefit from the openings or thinned forests created by timber cutting because of the post-cutting increase in low-growing browse and herbaceous vegetation. These species include elk *(Cervus canadensis),* moose, white-tailed deer, and mule deer *(Odocoileus hemionus;* Leopold 1966). Many studies document the increased use of logged forests, even clearcut areas, by these species (Patton 1976; Ffolliott et al. 1977; McNicol and Gilbert 1980). Large clearcuts which contain residual tree cover within the logged area, or small (20-40 acres) clearcut patches usually seem to provide better habitat than large expanses of clearcut forest (Short et al. 1977; Deschamp et al 1979; Lyon and Jensen 1980; McNicol and Gilbert 1980). The inhibition of forest regeneration by deer browsing (Crouch 1974; Kelty and Nyland 1981), moreover, seems to be greater on small rather than large clearcut areas (Mueggler and Bartos 1977). Thus, small clearings are superior deer habitat compared to large openings.

The benefits provided ungulates by clearcutting have been challenged. Davis (1977) criticized the often-made allegation that clearcutting, by replicating forest fires, has an equally-stimulating effect on numbers of deer and elk. He found that burns were used more than clearcuts, and he related the greater usage to the higher diversity of forage species and to the cover encouraged or provided by standing dead timber on burned plots. Severson and Kranz (1978) suggested that heavy cutting of bur oaks (Quercus macrocarpa) stimulates stump sprouts and thus increases browse, but that the removal of mature trees results in the loss of acorn production, thereby making clearcutting detrimental on winter deer range. Other studies, moreover, have found that slash inhibits use of clearcuts by deer and elk (Ffolliott et al. 1977; Lyon and Jensen 1980).

A most interesting questioning of the benefits of clearcutting for deer comes from Wallmo and Schoen (1980). They found that in the coastal forests of southeastern Alaska, the habitat most heavily used by deer is the mature, all-aged forest. This vegetation has an irregular canopy which permits light penetration to the forest floor, and thus the growth of low-growing forage plants. The canopy also intercepts snowfall and inhibits deep accumulations of which might otherwise bury the ground vegetation. Recent clearcuts have an abundance of forage, although winter snowpacks preclude their year-round use (see also Bloom 1978), and logging slash inhibits deer movements. More importantly, the forest regrowth on clearcuts is even-aged, which creates an unbroken canopy that prevents growth of forage plants. This condition of little food supply extends from the time of canopy closure, perhaps thirty years after logging, for as long as two centuries, when the even-aged characteristic of the forest has been modified by subsequent death and recruitment of trees. With clearcut rotations of a century, therefore, these forests will provide only poor deer habitat (Figure 17). Their conclusions are a direct challenge to the conventional wisdom that deer are favored by disturbance, and suggest that this interpretation resulted from the abundance of even-aged forests in North America:

> ... Most of the natural timber stands were even-aged because
> "disasters" such as disease, insect attacks, or fire kept returning
> the forest to early stages of succession ... Our early impressions of
> the relatively high habitat quality of early-seral shrub-herb vegeta-
> tion were based on comparisons with the poor quality of even-aged
> "oldgrowth" timber stands. Much of the uneven-aged overmature
> forest in North America was logged or burned before wildlife
> biologists had opportunity to record its productivity as deer habitat
> (Wallmo and Schoen 1980:459).

Wallmo and Schoen implied that the principle is generally applicable to all forests of the continent, but elsewhere Wallmo (1981:439) states that clearcutting is detrimental to deer populations only in the forests of the Northwest coast.

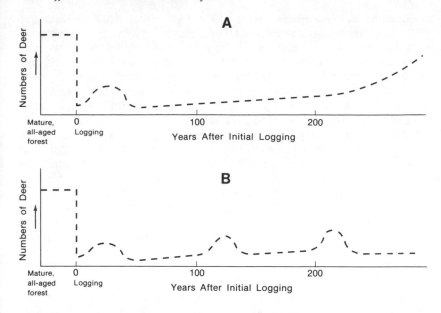

FIGURE 17   HYPOTHETICAL CHANGES IN ABILITY OF LOGGED
FORESTS OF COASTAL ALASKA TO SUPPORT DEER. **A.** One episode
of logging. **B.** Cutting on a 100-year rotation (Wallmo and Schoen 1980;
reproduced by permission of Society of American Foresters).

Small mammals. Like the large ungulates, small mammals respond differently to
the major disturbance of logging. Species that depend upon closed forests, such as tree
squirrels (Wolff and Zasada 1975), marten *(Martes americana;* Soutiere 1979), or
certain ground dwelling rodents (Geier and Best 1980), may decline with heavy cutting.
Still other animals that prefer open habitats may increase (Sims and Buckner 1973;
Geier and Best 1980). Group selection or patch clearcutting may often create more
favorable conditions than large clearcuts for some species (Patton 1977; Conroy *et al.*
1979).

Kirkland's study (1977) of the response of small mammals to clearcutting in West
Virginia is particularly interesting because he related feeding habits to changing environmental characteristics. The abundance of insectivores, least directly dependent
upon vegetation, varied little with clearcutting or reforestation after logging. Species
with generalist diets, granivore-omnivores, also showed little response. Grazers, however, increased greatly immediately after clearcutting because of the abundance of
herbaceous vegetation, but declined as woody plants again became dominant. Overall,
numbers and diversity of small mammals increased with clearcutting.

Birds. The species-specific response of birds to logging is well established (Hamilton and Noble 1975; Thomas *et al.* 1975; Titterington *et al.* 1979). Depending upon the
habitat requirements of different species for food, cover, and nesting, and the specific
silivicultural system employed, bird populations will reflect the altered vegetation
characteristics produced by timber harvesting.

Studies which evaluate birds in categories that represent biological functions such
as feeding are particularly revealing. Franzreb and Ohmart (1978), evaluating the

effects of an individual selection cutting in mixed conifers of Arizona, found that species which forage for insects on the barks or in the foliage of trees declined, but that several other "feeding guilds" increased. Those favorably influenced by the thinned forest included screeners, birds such as swifts that soar to catch flying insects; salliers, birds such as flycatchers that "sally forth" from perches to catch insects; ground feeders, such as robins; and low shrub/slash feeders, such as wrens. Overall, the number of species was not influenced by logging, but the number of individuals was greatly reduced.

Szaro and Balda (1979) found that birds in different feeding guilds varied in their responses to timber cutting depending upon the degree of thinning in ponderosa pine forests. Hammerers and tearers (woodpeckers) declined only in clearcuts; screeners, salliers, and ground feeders showed highest densities in shelterwood cuts; and foliage gleaners were most numerous in forests harvested by individual selection. Species diversity was highest in the shelterwood cut, and the number of individuals was highest in the selection cut. Only the heaviest cuts had fewer species and numbers than the uncut forest.

The importance to birds of individual forest characteristics, and the modification or loss of these features with logging, are often evaluated. Snags, dead but standing trees, are particularly important to cavity nesting birds, and timber-cutting procedures which not only leave dead trees standing but also allow snags to develop are frequently recommended (McClelland et al. 1979; Cunningham et al. 1980). Ecotones between clearings and forests are usually characterized by high numbers of species and individuals (Strelke and Dickson 1980), a factor to be considered in the determination of size and shape of clearings produced by logging (Thomas 1979). On the other hand, some species such as the warblers may suffer from small-scale patchiness in forest cover (Howe and Jones 1977; Whitcomb 1977). Riparian habitats probably suffer declines in birds even with light cutting (Stauffer and Best 1980). Intensively-managed forests, in which the variety in tree species and life forms is reduced in favor of uniform coverage by economically-useful trees, have fewer bird species than more "natural" forests (Dickson and Segelquist 1979).

A large expanse of forest being managed with clearcutting on a relatively short rotation will have much acreage in tree stands of dense saplings and poles. Studies have shown that immature forests support smaller numbers of individuals and species of birds than either younger stands or mature forests (Conner and Adkisson 1975; Austin and Perry 1979), probably because the tightly closed canopy of young forests prevents growth of low vegetation. Thus, the collective impact of many clearcut stands may be greater than that indicated by studies of individual clearcuts. Birds that depend upon old, "overmature" forest would be particularly vulnerable to such a management scheme.

Other Causes of Forest Disturbance.

Timber cutting is only one source of human-caused disturbance in forest, but the general effects of other causes of disturbance on wild animals are probably similar to those resulting from logging. Cleared right-of-ways may result in higher bird diversity because of the increases in ecotone-inhabiting species, although those dependent upon undisturbed forests may decline (Ferris 1979). In contrast, cleared chaparral had fewer species of small terrestrial vertebrates than undisturbed brush (Lillywhite 1977), and cleared pinyon-juniper woodlands similarly experienced decreased species diversity (O'Meara et al. 1981). Highways may, of course, introduce noise, gases, or other features which discourage birds directly rather than indirectly through vegetation (Van der Zande et al. 1980). The changing mix of birds associated with the recovering vegetation of strip-mined land seems to parallel the birds encountered on forest stands as they develop after clearcutting (Karr 1968). The same relationship has also been documented for small mammals (Sly 1976).

Livestock Grazing

Most evaluations of the interactions between wild animals and domestic livestock have focused on potential competition for forage (Stevens 1966; Smith et al. 1979; Lucich and Hansen 1981) rather than on assessment of the importance of livestock-altered vegetation as wildlife habitat. Nonetheless, situations in which livestock use has changed the structure of vegetation will also have altered wild animal populations. Leopold (1959), for example, suggested that increases in the density of shrubs in the sagebrush-grass vegetation over the last 100 years may have improved winter range for deer but decreased carrying capacity for pronghorn antelope (Antilocarpa americana); the vegetation change in part reflects heavy grazing by cattle and sheep. The same increase in shrubs may similarly benefit the bush-nesting Brewer's Sparrow (Spizella breweri), but not the grass and ground-nesting Vesper's sparrow (Pooecetes gramineus—Vale 1974), although Reynolds and Trost (1981) found that sheep grazing did not affect breeding birds in sagebrush. Reynolds (1979) determined that even lizard species in the sagebrush region respond to the altered plant cover, as suggested by the higher population of short-horned lizards (Phrynosoma douglassi) and smaller population of sagebrush lizards (Sceloporus graciosus) on grazed as compared to ungrazed plots. Similar changes in wildlife have likely resulted in other vegetation types in which woody plants have increased with livestock use.

A reduction in herbaceous vegetation cover, as might occur with heavy livestock grazing, may influence populations of wild animals. In a predictive model relating habitat characteristics and abundance of animal species, for example, Geier and Best (1980) found that most species of small mammals decreased with reduction in grass coverage. Heavy livestock use diminished both the height and coverage of non-woody vegetation on shorelines of ponds, moreover, with deleterious effects on waterfowl nesting (Whyte and Cain 1981). Grazing heavy enough to increase annual grasses and forbs in grasslands also creates conditions which encourage prairie dogs (Cynomys; Koford 1958).

Altered Fire Regimes

Most of the literature relating wild animals with fire reports the effects of individual fires upon wildlife (Krefting and Ahlgren 1974; Buech et al. 1977) or evaluates the importance of burning to certain animal species (Best 1979; Peek et al. 1979; Keay and Peck 1980; Halford 1981; Ream 1981). The effects of fire suppression are less commonly studied.

Ungulates that prefer disturbed vegetation suffer after suppression of fires. For example, Gruell (1979, 1980b) has argued that elk habitat in Wyoming has declined since elimination of fires which formerly encouraged the growth of herbaceous vegetation and deciduous woody plants. Forest fires in northern Idaho around the turn of the century, moreover, created vast areas of brush and young trees, prime habitat for elk. Numbers of elk increased dramatically and remained high until the mid-1950's. The recovery of the forests, under fire suppression policies, has been judged responsible for the decline in elk populations (Trout and Leege 1971). In addition, Leopold and Darling (1955) found that large uncontrolled fires in the boreal forest of Alaska improved forage supplies for moose, but reduced the carrying capacity on winter range for caribou. Numbers of caribou have decreased, whereas the moose has expanded its range across much of interior Alaska. Unlike caribou, winter range for bighorn sheep in Idaho may be improved by burning (Peek *et al.* 1979), and mule deer preferred burned range over unburned vegetation in Idaho (Keay and Peek 1980).

In general, animal species associated with undisturbed vegetation should be favored by fire suppression policies. This assessment is an oversimplification, however. A species that prefers mature forests, but does not necessarily gain from decreased fires, for example, is the marten. Infrequent fires are likely to be crown fires which cover large areas, effectively eliminating marten habitat. More frequent burns, on the other hand, may create conditions less than ideal but still tolerated by marten. Thus, fire suppression produces full mature forests, habitat that supports higher marten populations at one time. But over long time periods which encompass crown fire events, the forest system is characterized by brush vegetation or young forest that supports low populations of marten (Koehler *et al.* 1975; Koehler and Hornocker 1977).

New Equilibria

Wildlife species and numbers are strongly linked to characteristics of the vegetation. Thus, an equilibrium plant cover should have a mix of associated animals, perhaps varying spatially and temporally according to cycles of change within the vegetation. Human alteration of vegetation, particularly if viewed as creating new equilibria, will produce conditions more favorable for some animals and less for others, creating a new balance in wildlife characteristics appropriate for the new vegetation. The processes of major disturbance discussed are particularly important in this regard, because they involve large areas of landscape, as do strip-mining and purposeful manipulation. Other human activities may influence wildlife locally but sometimes dramatically, *e.g.*, offroad vehicles may reduce population of reptiles and burrowing rodents (Liddle and Scorgie 1981). Whether such wildlife changes are interpreted as good or bad depends upon one's view of the proper uses of the environment.

Effects on Physical Components of Environment

Vegetation influences not only the wildlife but also the physical characteristics of environmental systems. The plant cover, by interception and by affecting the movement of water from slopes to stream channels, alters the erosion on slopes and the discharge of streams. In addition, vegetation functions in the cycling of nutrients in the plant-soil system. Altered vegetation resulting from human activities may change, therefore, the rates and characteristics of surface erosion, the hydrologic characteristics of streams, and the availability of nutrients in ecosystems.

Surface Erosion

A reduction in the vegetation cover may, through decreased protection to the mineral soil, accelerate erosion by running water or mass-wasting. Disturbed plant covers, such as those altered by human activities, will be reduced in stature. Thus, for at least a period of time prior to vegetation regrowth, disturbance contributes to increased erosion.

Erosion potential may be represented by a diagram of the impact of logging upon erosion, which suggests that both gullying and mass-wasting may increase with timber-cutting (Figure 18). The cause and effect relationships, as represented by arrows, vary in intensity depending upon the nature of both the logging operation and the environment conditions. Swanston and Swanson (1976), for example, concluded that erosion in the Pacific Northwest by debris avalanches is increased to to four times by the reduction in vegetation stature associated with clear cutting, and 25 to 340 times by road construction. Although roads occupy only a small portion of the total area within a logged watershed, accelerated erosion linked to roads is about one-half of the total increase in erosion caused by timber harvesting operations. The disproportionate importance of the road system as a source of sediment is commonly recognized (Megahan and Kidd 1972; Anderson 1974; Swanson and Dyrness 1975; Meeuwig and Packer 1976; Anderson *et al.* 1976). The great range in the degree of increased erosion cited by Swanston and Swanson (1976) reflects not only road design and construction but also the natural stabilities of different landscapes. For example, the steep slopes and shallow, coarse-textured soils of the Idaho Batholith make that region particularly vulnerable to accelerated erosion (Megahan and Kidd 1972; Gray and Megahan 1981). On the other hand, an area of basalt flows in western Oregon experienced no increase

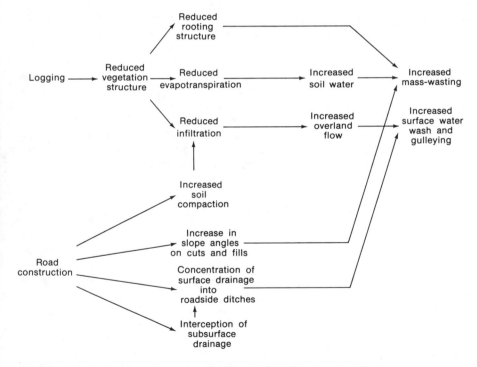

FIGURE 18   ACCELERATED EROSION CAUSED BY LOG-
GING. Relationships from Anderson *et al.* (1976). A similar but more
detailed model is given by Coates and Miller (1981).

in landsliding, except within road right-of-ways, even after clearcutting (Swanson and Dyrness 1975). Timber removal operations which involve use of machinery, such as tractor clearcuts with mechanical piling and burning of slash, produce lower infiltration rates and higher erosion rates than harvesting techniques, such as systems that lift and transport logs through the air, which minimize disturbance to the soil surface (Johnson and Beschta 1980). The length of time during which accelerated erosion is experienced will vary, but Swanson and Dyrness (1975) suggested that surface erosion rates may return to pre-logging levels in less than fifty years after clearcutting in unstable terrain in Western Oregon. Rice and collaborators (1979) concluded that with a cutting rotation of 50 years in mountainous northwestern California, the accelerated erosion after each timber harvest would not result in complete soil loss for 8,000 to 34,000 years.

Other human activities which, like logging, reduce the stature of the plant cover and/or compact the soil may lead to accelerated erosion. These activities include clearing forests for agriculture (Knox 1977; Hill 1976), grazing (Branson et al. 1972; McGinty et al. 1978), trampling (Helgath 1975; Bratton et al. 1979), off-road vehicles (Vollmer et al. 1977; Eckert et al. 1979), and conversion of woody vegetation to herbaceous plant cover for better forage production (Pitt et al. 1978). Forest fires that reduce the stature of vegetation are often followed by a period of accelerated erosion (Meghan and Molitor 1975; Anderson et al. 1976). Conversely, human activities that produce an increased plant cover, such as reforestation of abandoned farmland, should result in decreased erosion (Anderson et al. 1976). Generalization about the long-term effects of fire suppression on surface erosion is complicated because increases in forest density, associated with reduction in fire frequency, may decrease erosion rates, but increases in intensity of burns, associated with accumulations of fuels, may accelerate erosion compared to more frequently burned forests.

Streamflow

Variations in streamflow in a stable watershed reflect the pathways by which water moves from slopes to stream channels. Vegetation influences amount of water delivered to streams, as it determines interception and transpiration water loss from watersheds. Vegetation, by influencing infiltration rates, also affects the speed by which water reaches stream channels and thus the discharge response to precipitation events. Changes in vegetation associated with human activities may thus alter both annual water yield and peak discharge flows (Figure 19).

Annual water yield. In general, the deeper and the denser the rooting of vegetation, the greater the transpiration, and the smaller the water yield of a watershed. Decreases in the plant cover characteristically result in an increase in water yield (Dunne and Leopold 1978). Logging serves as an example. Increases in streamflow are typically greater with heavier timber cuts, and they are potentially larger in humid than in drier areas; the persistence of higher flows is inversely related to the rapidity and vigor of forest regrowth (Harr et al. 1975; Harr 1976; Patric and Aubertin 1977; Lima et al. 1978; good summaries are provided by Anderson et al. 1976, and Dunne and Leopold 1978). Logging may not always increase water yield or low flows, however, as indicated by Harr (1980). Fires that kill trees and other vegetation are often followed, like logging, by increased water yields (Helvey et al. 1976; Anderson 1976; Anderson et al. 1976); regrowth of the forest will result in declining stream flow (Jaynes 1978). Livestock grazing seems not to influence annual water yields, presumable because it does not strongly effect transpiration rates (Meeuwig and Packer 1976; Anderson et al. 1976;

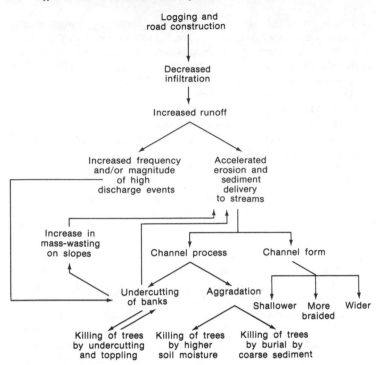

FIGURE 19   RESPONSE OF STREAMS TO CLEARCUT LOGGING IN COAST REDWOOD FORESTS OF NORTHWEST CALIFORNIA (Vale 1978).

Branson *et al.* 1972). Apparently, the effects of mining on water yields are highly variable (Meeuwig and Packer 1976). Purposeful manipulation of vegetation to increase water yields for greater water supplies is often evaluated, particularly for western rivers such as the Colorado (Hibbert 1979). In contrast, the cutting of the forests in the Amazon may increase runoff, but in this situation the results are usually considered deleterious (Gentry and Lopez-Parodi 1980).

Peak flows. The effects of vegetation disturbance, such as logging, on peak flows in streams are less consistent than the general increase in water yield. Moreover, Harr (1976) suggested that the larger the discharge, the less the importance of cutover forest as a contributor to a high streamflow event. Nonetheless, where logging is associated with widespread soil disturbance, particularly with road construction, infiltration can be lessened, and storm runoff can be greatly increased. The western states, where growing season precipitation is characteristically low (thereby slowing reforestation), where large trees make heavy harvesting machinery more common, and where dissected topography allows for rapid response of streams to precipitation, the problem of increased high flows with logging is particularly troublesome (Anderson *et al.* 1976).

Other vegetation changes that result in reduced vegetation cover and/or increased soil compaction produce higher peak flows in streams. These changes include hot, widespread forest fires; livestock grazing; and construction activity (Branson *et al.*

1972; Meeuwig and Packer 1976; Anderson *et al.* 1976; Helvey *et al.* 1976). Curtis (1977) argued, however, that mining in a Kentucky watershed reduced peak flows, perhaps because of higher infiltration and greater basin storage capacity provided by rock broken during the mining.

Soils

The strong link between vegetation and soil means that vegetation altered by human activities will almost necesarily have corresponding changes in soils. These changes include reduction in organic matter caused by grazing (Johnston *et al.* 1971), creation or movement of non-wettable layers caused by fire (Scholl 1975; Dyrness 1976; DeBano 1981), altered populations of microorganisms casued by logging (Jurgensen *et al.* 1979), or exposure of "new" initial ("parent") materials by mining (Burton *et al.* 1978). Nutrient concentrations may be influenced if surface erosion or infiltration rates are altered, whether by logging (Corbett *et al.* 1978), fire (DeByle 1976; DeBano *et al.* 1979; Wells *et al.* 1979), or clearing for agriculture (Hill 1976).

Does timber-cutting have long-term effects on soil fertility? Bormann and Likens (1979) concluded that clearcutting on a rotation of at least 65 years will allow mainte-nance of soils in the White Mountains of New England. After reviewing the literature, Bockheim and collaborators *(n.d.)* concluded, "In general, nutrient losses following clearcutting are small to negligible, particularly where small patches or strips are removed." Kimmins (1977) made a similar statement about the effects of whole-tree harvesting upon available nutrients; he stressed, however, that short timber-cutting rotations may produce gradually deteriorating soils. On the other hand, Vitousek and collaborators (1979) concluded that forests in the United States seemed generally vulnerable to loss of nitrates after disturbance. In any event, strong economic pressures for logging marginal sites where regrowth is slow and soil erosion potential is high, or for cutting too much and too often are much more likely to result in soil degradation than more conservative logging programs.

Changes in the Physical Characteristics of Environment

Hydrologic and edaphic characteristics of environments reflect the nature of vege-tation. The viewpoint of integrated factors influencing hydrology underlies such com-mon expressions as the universal soil loss equation and curve numbers for estimating runoff from precipitation events. We can think of particular features of streamflow, erosion, and soil, then, as associated with a particular plant cover in a kind of equilib-rium. A change in vegetation may force an adjustment in hydrologic and edaphic characteristics, just as it induces change in the nature of wildlife.

5

Studying Vegetation Change

Studying vegetation change caused by human activities is little different in approach from evaluations of vegetation change initiated by natural causes. Research designs and sampling procedures in plant ecology are useful, and most common techniques can be employed given appropriate situations. Certain procedures, however, seem particularly fruitful in the evaluation of human impacts on plant cover, and they are utilized in many of the studies cited.

Historical Sources

Materials such as written descriptions, land survey records, and photographs often provide valuable insight about past vegetation conditions which may then be compared with contemporary plant covers. These sources lack the objectivity which scholars typically demand of their data, but if used cautiously (determining the general characteristics of vegetation rather than precise measures) historical sources frequently prove invaluable.

Written Descriptions

Journals, diaries, letters, and reports often contain mention of vegetation conditions. Such sources are particularly common and available for the period of European expansion westward across North America. The use of written descriptions as evidence for past landscapes is constrained by the vagueness of writing compared to numerical data, the biases of the writers who may have passed on incorrect impressions about the vegetation, and the misconceptions of contemporary researchers who may read their own biases into the writings. The problems stemming from the ambiguities inherent in these sources can be minimized by some rules of thumb: Use as many different descriptions as possible, rather than depend upon one or a few; interpret the writings to gain impressions of dominant life forms, but not to distinguish between species of similar life form; and give little credence to vague words such as "barren" or "luxuriant". When followed, these procedures allow use of written descriptions to reconstruct past vegetation, as exemplified by Vale (1975a) in the sagebrush-grass area of the interior West, Curtis (1959) in the prairies of Wisconsin, Thompson (1961) in the riparian forests of California's Central Valley, Leopold (1951) in the desert southwest, Johannessen (1971) in the Willamette Valley of Oregon, and Russell (1981) in the forests of New Jersey). Any major collection of historical writings, whether in public or private libraries, historical socieities, or government offices, is a suitable place to search for descriptions of early vegetation.

Land Survey Records

The initial land surveys in the township and range grid system in the United States were accomplished by individuals who walked the section lines, recorded trees that they encountered on the line, described major characteristics of the plant cover, and marked the corners of the sections with posts or piles of rock. Most importantly for vegetation studies, the surveyors located at each corner "witness" or bearing trees, which were presumably the trees closest to the corner in each (often only two) of the four quadrants formed by the survey lines. The size and species of the witness trees, as well as the distance between each tree and the corner, was recorded in a notebook. The resulting information is thus a sample, uniformly distributed but quantitative, of the trees within a forest which may be used to reconstruct such ecological characteristics as species composition, size structure, density of stand, and many other descriptive features of early forests (Figure 20). Questions about possible biases of surveyors, who may have selected, for example, trees of certain species or sizes as bearing trees, have been examined by Bourdo (1956), who suggested simple tests for bias which may be applied to the data.

Land survey records have been widely used to reconstruct past forest vegetation in the eastern and midwestern states (Stearns 1949; Ward 1956; Johannessen et al. 1971; Rankin and Davis 1971; Siccama 1971; Delcourt and Delcourt 1974; Delcourt 1975; Lorimer 1977; Flower 1980; Russell 1981). From reconstructed data, maps at the state or county scale have been prepared (Trewartha 1940; Stroessner and Habeck 1966; Tans 1976; Finley 1976; Gross and Dick-Peddie 1979; Leitner and Jackson 1981). In part as a means to compare present with past conditions, Cottam and Curtis (1949) developed the point-quarter sampling method which duplicates some of the procedures used by the early surveyors, and which may be used in a contemporary stand to generate a data set that may be contrasted with that from the surveyors' notebooks. Cottam (1949) followed this procedure in a southern Wisconsin site and found that the fire-maintained savanna vegetation of pre-European times had been replaced by a closed forest with an accompanying change in species importance (Table 7). Land survey records are usually available in state land offices with duplicate sets sometimes kept in county seats or historical libraries. In the West, survey records are typically housed in the state offices of the Bureau of Land Management.

Photographs

Photographs of landscapes in the United States extend back more than a century and, thus, may record early vegetation conditions. Rephotographing the same views today permits the creation of photo pairs that graphically portray changes in the plant cover. Rarely did photographers record true presettlement conditions because their craft developed too late to do so. Nonetheless, photos are frequently old enough to permit analyses of vegetation change over long periods. As with the written descriptions, photographic comparisons have the obvious problem of yielding qualitative rather than quantitative information, and they are probably best used to determine general characteristics of plant cover, particularly life forms, rather than species composition. Time of year and short-term climatic variations may also introduce variability which could be misinterpreted. It is also important that the exact view be duplicated, preferable with a lens of identical length, because variations in vegetation from place to place may be mistakenly viewed as temporal changes. The use of such photos has

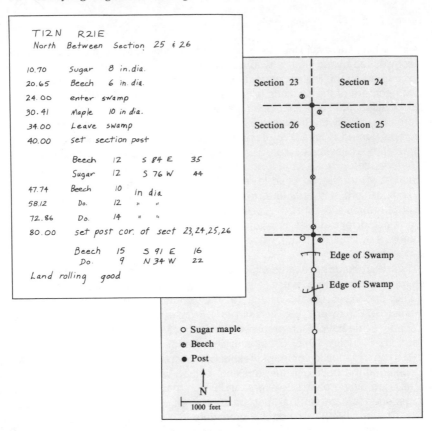

FIGURE 20   PAGE FROM LAND-SURVEYOR'S NOTEBOOK IN EASTERN WISCONSIN, WITH CORRESPONDING SKETCH MAP. The numbers in the left-most column record distance along section line from starting point, in chains (one chain = 66 feet or 20.12 meters). In the next column, tree species encountered on the line or used as bearing trees are identified. In the third column, tree diameter is recorded (inches). At the mid-point on the line (40 chains from the starting point) and at the final corner of the township (80 chains), a post was set and two bearing trees were selected. The diameter of each bearing tree is given, as well as its direction from the post and distance, in links (about 8 inches or 20.3 cm).

enjoyed considerable attention in recent years, particularly in the western states (Leopold 1951; U. S. Forest Service 1971; Gary and Currie 1977; Martin and Turner 1977; U.S. Bureau of Land Management 1976a, 1976b; Gruell 1980a; Rogers 1980). Some particularly impressive examples of this technique are Gibbens and Heady's (1964) examination of the forest in Yosemite Valley in California, and Hastings and Turner's study (1965) of the arid and semi-arid vegetation in the United States-Mexico border region.

Old photographs, whether in photo collections or in printed books, are available in a wide variety of places, including libraries, historical societies, and government offices.

The U.S. Geological Survey Photographic Library in Denver, Colorado, houses all of the pictures taken under the auspices of the survey, including those of the early western explorations (Vale 1973). Studies of vegetation change over the last few decades could easily incorporate photographic comparison, including aerial photographs, a source of information little used.

Ecological Sources

With appropriate research design, most ecological sampling techniques may be useful in questions of vegetation change. A comparative approach is frequently used in studies of human impacts on vegetation, in which a population altered by a human activity is compared to another population not so affected but otherwise in an equivalent setting. Procedures such as t-tests, chi-square analysis, or analysis of variance, testing the significance of the differences between populations, are frequently useful. Sometimes populations compared may be discrete, such as a pipeline right-of-way with its construction disturbance in otherwise undisturbed forest, or an ungrazed plot within generally grazed rangeland. In other situations, the degree of an impact may change continuously through space, toward sites not altered, such as pollution away from a smokestack or trampling away from a trial. In each case, the identification of the human impact depends upon isolating it as a variable, and determining whether or not the vegetation altered by people is significantly different from the vegetation not so influenced.

Relicts

Occasionally individual plants may give evidence of past vegetation characteristics. For example, large horizontally-extended branches of trees, a so-called "open-growth form", is usually interpreted as indicating growth in a non-forested environment; the form may persist, thus suggesting a former savanna, even if the fires which

TABLE 7   COMPARISON OF VEGETATION FROM 19th CENTURY LAND SURVEY AND 20th CENTURY VEGETATION FOR AN AREA IN SOUTHERN WISCONSIN

Data	1834	1946
Trees/acre	14.3	143
Basal area in square feet/acre	12.6	105.1
Average basal area in square feet per tree	0.91	0.83
Frequency[a] of common species		
Bur oak *Quercus macrocarpa*	72%	8%
White oak *Quercus alba*	37%	83%
Black oak *Quercus velutina*	20%	53%

[a]Frequency is the percent of samples which contained a given species.
Source: Cottam 1949: 283. Reproduced by permission of Ecological Society of America.

maintained the open vegetation are eliminated and a dense forest develops beneath the old trees (Daubenmire 1968). Relict distributions, such as herbaceous prairie plants growing in closed forests, may similarly indicate past vegetation conditions (Bray 1957).

Relict areas, as opposed to relict individuals, are invaluable as reference locales against which to assess such human impacts as logging or grazing (Moir 1972; Franklin 1977). Natural relicts, areas which have not been altered by human activities because of inaccessibility, may preserve true precolonial conditions. Politically-created natural areas usually protect landscapes only in partially pre-European conditions. Protected areas often are used to assess the trend in vegetation characteristics after certain human activities have been eliminated.

Several references identify relict areas: A directory of research natural areas on federal lands (Federal Committee on Research Natural Areas 1968), a complete description of the environments of research natural areas in Oregon and Washington (Franklin *et al.* 1972), reports on individual research natural areas issued by federal land agencies (Bjorkbom and Larson 1977), lists of grazing exclosures (small areas fenced off from livestock — Laycock 1969), many state catalogues of natural areas (Germain 1977; California Natural Areas Coordinating Council *n.d.*), and areas purchased by The Nature Conservancy and similar organizations. Units of the National Park System often serve the relict function, as do many state and local parks.

Repeated Sampling

Ecological studies have been made for a sufficiently long time that the sampling from early studies can often be repeated, thereby providing the basis for evaluations of vegetation change. The development of the vegetation in the time between sampling may reflect either the influence on human activities during the period or the recovery of the vegetation after elimination of such activities. This approach of repeated ecological sampling is used often, but warrants even greater attention (Humphrey and Mehrhoff 1958; Robertson 1971; Schmelz 1975; Phillips 1976; Abrell and Jackson 1977; McCormick and Platt 1980).

Size and Age Structures of Species Populations

A frequently-used technique for identifying vegetation change of individuals is size or age cohorts. Sizes are usually expressed as main stem diameter at breast height. Ages are determined with increment borings, small cylinders of wood extracted from, and presenting a cross section of, the main stem from which annual growth rings are read. The resulting size or age structures are interpreted as representing the dynamics of the species populations (Leak 1975). A structure with progressive decreasing numbers in ever larger or older individuals is called an inverse J-shaped population because of the shape of the graph produced by such a structure (Figure 21). Such a population structure is usually considered "balanced," one just maintaining itself (Meyer 1952; Daubenmire 1968). Few small or young individuals, in contrast, suggest a species not successfully reproducing (Figure 21), and an abundance of small or young may indicate an "invasive" species. Size and age structures have been used to identify changing reproductive success associated with climate fluctuation (Leak and Graber 1974), windthrow events (Lorimer 1980), livestock grazing (Niering *et al.* 1963), timber cutting (Leak 1964; Johnson and Bell 1975), or fire suppression (Cooper 1960; McBride

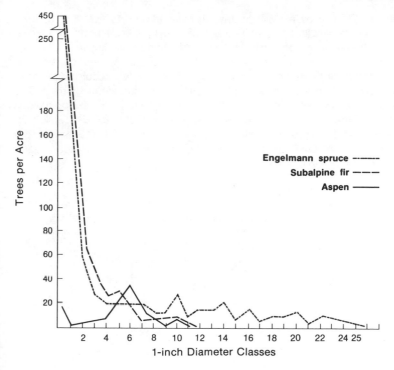

FIGURE 21 SIZE STRUCTURES OF THREE TREE SPECIES IN UTAH. Engelmann spruce and subalpine fir both have the inverse J-shaped structures that represent stable, reproducing populations. Aspen has the inverted U-shaped structure of a non-reproducing population (Hanley *et al.* 1975).

1974; Vale 1977). Age or size structure interpretations require knowledge of the natural history of the species involved, and simple comparisons with an ideal inverse J-shaped curve can lead to faulty conclusions. Causes of deviations from the J-shaped form, many of which are the natural causes of fluctuations in numbers mentioned in the first chapter, do not necessarily indicate "unstable" populations (Figure 22). For example, a species like eastern hemlock may not reproduce continuously, and thus at any one time its population structure may have few small or young trees. Given more time, though, reproduction will occur, and the population would be seen to be "stable."

Soil Characteristics

The processes linking vegetation and soil often mean that certain characteristics of one may be associated with the other. Given an environmental change, the plant cover may respond more rapidly than some soil features, with the soil retaining evidence of the former vegetation (Birkeland 1974:197-210). Usually used in studies of climatic fluctuations, soil characteristics may also prove useful in evaluations of human alteration of vegetation. With tree invasion of grasslands or savannas, as might occur with elimination of fires, available nitrogen and base saturation in the upper soil will usually decrease rather rapidly. The low mobility of clay in grasslands and its increased mobility in forests, in contrast, means that an enriched clay content high in the profile of a forest soil may indicate the presence of a former grassland (White and Ricken 1955).

The soils on either side of forest-grassland ecotones also have been shown to be different in the content of biogenic opal, or opal phytoliths, microscopic silica compounds secreted by plant cells and produced in particular abundance by grasses. The forms of phytoliths vary with plant species, and thus distinctive phytolith populations accumulate under different vegetation. Because it is extremely stable, opal persists for long periods even with changes in vegetation. Thus, former grasslands may be detected long after other soil characteristics have adjusted to the non-grassland vegetation (Jones and Beavers 1964; Miles and Singleton 1975).

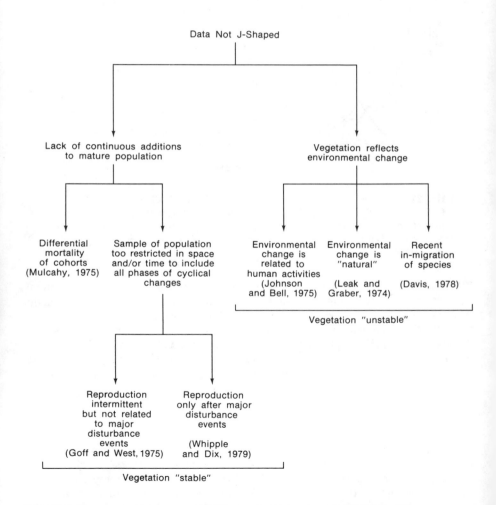

FIGURE 22   POSSIBLE INTERPRETATIONS OF A SIZE OR AGE STRUCTURE OF A TREE SPECIES THAT IS NOT INVERSELY J-SHAPED

Pollen

Wind-pollinated species produce large amounts of pollen that are easily dispersed in the air. Where the pollen accumulates and is preserved in acid bogs, lake sediments, or sometimes soil, evidence of past vegetation is preserved. Pollen analysis is a fundamental technique of researchers who reconstruct past climates, drawing upon the strong association between certain plant species and climatic characteristics. Determining the vegetation and thus the climate of the Holocene in North America is an inquiry that has received considerable attention (Davis 1976; Bernabo and Webb 1977; Kapp 1977; Davis 1978; Delcourt and Delcourt 1979; Van Devender and Spaulding 1979; Wells 1979; Delcourt et al. 1980). The pollen record has also been used to document vegetation change caused by human activities, as, for example, forest clearing for agriculture (Bernabo and Webb 1977; Zant 1979), logging (Davis 1973; Swain 1980), or heavy grazing by livestock (Davis et al. 1977). Explaining the causes of changes in a pollen core is sometimes complicated, however, by a multiplicity of types of change in vegetation, even in an equilibrium condition.

Combining Sources

Rarely can a researcher bring all of the various sources of information about vegetation change to bear on a particular problem. But all to often only one source is utilized. Historical materials may provide a general impression about vegetation that appeals to our desire to understand broad patterns of change. Ecological approaches and data, on the other hand, enable more precise evaluations, particularly important in situations where the change is subtle. Brought together, the two major sources can combine to produce studies that are rich and fun, as well as accurate and enlightening.

6

Vegetation Change and Human Purpose

Many people, whether general citizen or environmental specialist, are accustomed to thinking that "natural" vegetation and ecosystems are preferable to the altered conditions resulting from human activities. Most of us thought of fires as destructive until science indicated that they were natural, and thus a process to be admired. Logging, particularly heavy cutting, is frequently decried as harmful, as is intensive livestock grazing. Few people have anything positive to say about the effects on wild vegetation of off-road vehicles, strip-mining, or air pollution.

Yet, as was discussed in earlier chapters, it is difficult to generalize about the ecological effects of humans on vegetation. Some activities decrease plant diversity, while others increase it. Some cause reduced productivity, but some create more productive systems. Some may result in a more varied vegetation structure, although many reduce the variety of the dominant life forms. The effects of altered vegetation on wildlife, hydrology, and soils are similarly variable depending on the human activity and the particular environmental situation.

In the absence of consistent, predictable ecological measures, human alteration of vegetation is usually evaluated by assessing the likelihood of vegetation recovery to a pre-impact condition. Some human activities may change an ecological characteristic in one direction while other activities may have an opposite effect. Nevertheless, our reaction to all activities is to ask, how well the features of the plant cover existing prior to human alteration will become reestablished. Critics of clearcut logging, for example, assert that such timber cutting eliminates the forest, or at best replaces large old trees with small young ones. In either case, the essence of the criticism is that a new equilibrium has replaced the old. Similarly, heavy grazing alters the vegetation in pastures and rangelands. If such areas are not grazed excessively, we imagine that the most palatable species will recover, but if they are overgrazed, we perceive the establishment and presistence of a new plant cover, a new equilibrium. In fact, such a plant cover becomes the measure of overgrazing. Individual fires were formerly seen as creating new vegetation conditions, but today fire suppression is recognized as doing so. Disturbance associated with construction, trampling, off-road vehicles, or pollution each may be deemed acceptable, but only if the post-disturbance development of the plant cover can recreate the predisturbance vegetation. If human activity occurs in an environment characterized by little or slow rates of vegetation development, or if the plant cover is purposefully manipulated to create and maintain a new equilibrium, the label of "destruction" is usually applied. What existed, or what was perceived to exist, before the alteration is the standard against which the impact of the human activity is evaluated.

Several factors complicate the identification of the replacement of one equilibrium by another. First, the perception of an equilibrium condition, or the opposite, a vegetation change, has an implied time scale. For example, a clearcut plot that is dominated by brush and only a few trees may be viewed as a deforested area. But a thick cover of trees may gradually develop. Over a long time scale, then, the clearcut may be interpreted as but a temporarily disturbed state in a continuing equilibrium. Second, in some situations the spatial scale is an important consideration. A strip of trampled ground beside a trail, for example, typically has reduced plant diversity compared to similar but untrampled areas away from the trail. For larger areas of terrain, however, the trampled strip may influence diversity no more than the "naturally" disturbed zone along a dry ravine. Thus, at a larger scale the equilibrium diversity seems unaffected by the trampling.

Our degree of analytical generality also influences whether we interpret human activity as causing vegetation change, creating a new equilibrium. Livestock grazing of increasing intensity, for example, may alter the relative importance of different grass species in a grassland, then permit invasion by shrubs to produce a mixture of grasses and brush, and finally allow much exposure of bare soil. If we perceived the pre-grazed vegetation as simply a grassland without considering its composition, only the more severely altered plant cover would be seen as a changed vegetation. On the other hand, if we demanded exact maintenance of pregrazing species composition, any grazing, however light, would initiate change.

Finally, and perhaps underlying the first three points, the degree to which natural change is incorporated into our perception of the vegetation system plays a critical role in the evaluation of human impacts (Figure 23). If we downplay the importance of change in our definition of vegetation (the extreme position being a static view of the plant cover), any human impact would necessarily be disruptive. On the other hand, if we incorporate natural change as part of our system, human alteration will seem less an aberration. The former view would see one of the smaller boxes in Figure 23 as describing vegetation, perhaps at most allowing for the dynamics of the larger horizontal box in the upper third of the diagram. That portion is parallel to the upper level of Figure 7 (page 15). The latter view, incorporating a wide degree of disturbance as inherent in vegetation, is symbolized by the largest boundary in the diagram, incorporating multiple phases of disturbance, again as in Figure 7. Over long time scales which extend into the Pleistocene and beyond, and over large spatial scales which encompass large portions of continents, the composition of vegetation has changed so much that many contemporary human impacts seem insignificant.

Although we recognize that the perceived importance of human alteration of vegetation is relative to the ecological system of reference from which change is viewed, we need not dismiss the significance of such alteration to individuals, groups, or socieities. Human-induced vegetation change, of whatever scale, may be deleterious if that change is contrary to the purposes envisioned for a landscape. The assigning of value and meaning to vegetation change depends upon the identification of human goals. For most of us, the goal of maintaining a world that is biologically productive and diverse is so fundamental that vegetation change which seems to threaten that productivity or diversity is easily condemned. Examples include the decrease in palatable plants on rangeland caused by heavy grazing, the elimination of forests by air pollution, or the accelerated erosion associated with abusive strip-mining. On a world scale, still other examples come to mind — deforestation and increased erosion in Third-World

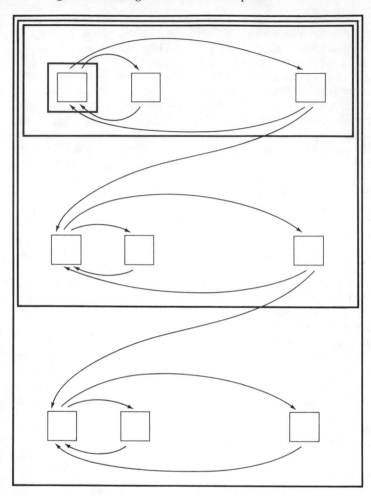

FIGURE 23   DIFFERENT BOUNDARIES FOR THE DEFINI-
TION OF "VEGETATION" OR "VEGETATION SYSTEM." The
diagram illustrates alternative conceptions of vegetation based on
the degree of disturbance inherent in the definition. For details of the
components and processes, see Figure 7, page 15.

countries where fire wood is an essential resource, or elimination of arid land vegetation
caused by excessive grazing.

Other human-induced vegetation change, however, seems less necessarily to
represent a threat to the biological productivity of Earth. In these instances, different
human goals become the basis for evaluating vegetation change. Consider, for example, an area of old-growth forest in western North America or in a tropical country. If
someone felt that the area should be used for food production, permanent clearing of
the land would be defined as "good." In contrast, a forester interested in using the same
area for producing wood fiber or pulp would likely advocate heavy cutting, but not
permanent clearing, in order to increase the forest's productivity. Another forester

concerned with the production of high-grade lumber might feel that a selective logging system to be "best." If the landscape were dedicated to watershed management, specifically reducing peak flows in downstream locations, a policy of light and infrequent cutting might be the only disturbance considered "desirable." If the area was designated as a park, wilderness, or sacred reserve, no cutting would be appropriate. Thus, human values, not the ecological effects *per se*, determine the "goodness" or "badness" of human alteration of old-growth forest. Observers might agree that obliteration of all vegetation is undesirable, but arrival of the miner might shatter even that consensus. Distinctions between good use and bad use, destructive use and constructive use, conserving and exploiting, or preserving and harvesting become muted and unclear.

People who argue for more gentle uses of vegetation — modest timber cutting, light livestock grazing, prohibition of off-road vehicles, revegetation of strip-minded land, elimination of air pollution damage, and protection of wilderness — are no more (or less) emotional or subjective than others who argue for different uses of vegetation. Everyone interprets human alteration of vegetation with certain values, a certain vision (however cloudy or inconsistent) about a desirable world. Ecological assessments of human impacts help each of us, whether scientist or general citizen, decide whether to condone or condemn any particular impact. But how we chose to interpret vegetation altered by people is also subjective. Ultimately our determination of right or wrong comes more from our heart than from our head.

Bibliography

Abrell, D. B. and M. T. Jackson. 1977. "A Decade of Change in an Old-Growth Beech-Maple Forest in Indiana," *American Midland Naturalist* 98:22-32.

Ahlgren, C.E. 1974. "Effects of Fire on Temperate Forests: North Central United States," pp. 195-223 in Kozlowski and Ahlgren (1974).

Ahlgren, C.E. 1976. "Regeneration of Red Pine and White Pine Following Wildfire and Logging in Northeastern Minnesota," *Journal of Forestry 74:135-140.*

Alexander, R. R. and C. B. Edminster. 1980. *Management of Ponderosa Pine in Even-Aged Stands in the Southwest.* Rocky Mountain Forest and Range Experiment Station, Research Paper RM-225.

Alvarez, H. et al. 1974. "Factors Influencing Plant Colonization of Mine Dumps at Park City, Utah." *American Midland Naturalist* 92:1-11.

Amundson, D. C. and H. E. Wright. 1979. "Forest Changes in Minnesota at the End of the Pleistocene." *Ecological Monographs* 49:1-166.

Amundson, R. G. and L. H. Weinstein. 1980. "Effects of Airborne F on Forest Ecosystems," pp. 63-78 in *Effects of Air Pollutants on Mediteranean and Temperate Forest Ecosystems.* Pacific Southwest Forest and Range Experiment Station, General Technical Report PSW-43.

Anderson, H. W. 1974. "Sediment Deposition in Reservoirs Associated with Rural Roads, Forest Fires, and Catchment Attributes," pp. 87-95 in *Proceedings of Symposium on Man's Effects on Erosion and Sedimentation.* International Association of Hydrologic Science, Publication 113.

Anderson, H. W. 1976. "Fire Effects on Water Supply, Floods, and Sedimentation," *Proceedings,* Tall Timbers Fire Ecology Conference 15:249-260.

Anderson, H. W. et al. 1976. *Forests and Water: Effects of Forest Management on Floods, Sedimentation, and Water Supply.* Pacific Southwest Forest and Range Experiment Station, General Technical Report PSW-18.

Anderson, J. E. and K. E. Holte. 1981. "Vegetation Development Over 25 Years Without Grazing on Sagebrush-Dominated Rangeland in Southeastern Idaho," *Journal of Range Management* 34:25-29.

Anderson, R. C. and O. L. Loucks. 1979. "White-tail Deer Influence on Structure and Composition of *Tsuga canadensis* Forests," *Journal of Applied Ecology* 16:855-861.

Anderson, S. et al. 1977. "The Effect of Transmission-line Corridors on Bird Populations," *American Midland Naturalist* 97:216-221.

Antos, J. A. and J. R. Habeck. 1981. "Successional Development in *Abies grandis* Forests in the Swan Valley, Western Montana," *Northwest Science* 55:26-39.

Antos, J. A. and R. Shearer. 1980. *Vegetation Development on Disturbed Grand Fir Sites, Swan Valley, Northwestern Montana.* Intermountain Forest and Range Experiment Station, Research Paper INT-251.

Arno, S. F. 1976. *The Historical Role of Fire on the Bitterroot National Forest.* Intermountain Forest and Range Experiment Station. Research Paper INT-187.

Arno, S. F. 1980. "Forest Fire History in the Northern Rockies," *Journal of Forestry* 78:460-465.

Arno, S. F. and D. H. Davis. 1980. "Fire History of Western Red Cedar/Hemlock Forests in Northern Idaho," pp. 21-26 in M. A. Stokes and J. H. Dieterich (editors), *Proceedings of the Fire Ecology Workshop,* Rocky Mountain Forest and Range Experiment Station, General Technical Report RM-81.

Arno, S. F. and K. M. Sneck. 1977. *A Method for Determining Fire History in Coniferous Forests of the Mountain West.* Intermountain Forest and Range Experiment Station, General Technical Report INT-42.

Ashby, W. and G. Weaver. 1970. "Forests Regeneration on Two Old Fields in Southwestern Illinois," *American Midland Naturalist* 84:90-104.

Auclair, A. and G. Cottam. 1971. "Dynamics of Black Cherry in Southern Wisconsin Oak Forests," *Ecological Monographs* 41:153-177.

Austin, D. and M. Perry. 1979. "Birds in Six Communities Within a Lodgepole Pine Forest," *Journal of Forestry* 77:584-586.

Bailey, R. G. 1976. *Ecoregions of the United States.* Ogden, UT: U.S. Forest Service.

Baker, H. 1972. "Human Influences on Plant Evolution," *Economic Botany* 26:32-43.

Baker, H. 1974. "The Evolution of Weeds," *Annual Review of Ecology and Systematics* 5:1-24.

Barney, M. A. and N. C. Frischknecht. 1974. "Vegetation Changes Following Fire in the Pinyon-Juniper Type of West-Central Utah," *Journal of Range Management* 27:91-96.

Barrett, S. 1980. "Indians and Fire," *Western Wildlands* 6, 3:17-21.

Basile, J. V. 1979. *Elk-Aspen Relationships on a Prescribed Burn.* Intermountain Forest and Range Experiment Station, Research Note INT-271.

Batzli, G. O. and F. A. Pitelka. 1970. "Influence of Meadow Mice Populations on California Grassland," *Ecology* 51:1027-1039.

Bates, G. H. 1935. "The Vegetation of Footpaths, Sidewalks, Carttracks and Gateways," *Journal of Ecology* 23:470-487.

Bay, R. 1976. "Rehabilitation Potentials and Limitations of Surface Mined Lands," *Transactions, North American Wildlife and Natural Resources Conference* 41:345-355.

Bayfield, N. G. 1973. "Use and Deterioration of Some Scottish Hill Paths," *Journal of Applied Ecology* 10:635-644.

Beetle, A. 1974. "The Zootic Disclimax Concept," *Journal of Range Management* 27:30-32.

Bernabo, J. C. and T. Webb. 1977. "Changing Patterns in the Holocene Pollen Record of Northeastern North America: A Mapped Summary," *Quaternary Research* 8:64-96.

Best, L. 1979. "Effects of Fire on a Field Sparrow Population," *American Midland Naturalist* 101:434-442.

Betancourt, J. L. and T. R. VanDevender. 1981. "Holocene Vegetation in Chaco Canyon, New Mexico," *Science* 214:656-658.

Billings, W. D. and H. A. Mooney. 1959. "An Apparent Frost Hummock-Sorted Polygon Cycle in the Alpine Tundra of Wyoming," *Ecology* 40:16-20.

Birkeland, P. 1974. *Pedology, Weathering, and Geomorphological Research.* New York: Oxford University Press.

Biswell, H. H. 1963. "Research in Wildlife Fire Ecology in California," *Proceedings, Tall Timbers Fire Ecology Conference* 2:63-97.

Biswell, H. H. 1973. "Fire Ecology in Ponderosa Pine-Grassland," *Proceedings, Tall Timbers Fire Ecology Conference* 12:69-96.

Biswell, H. H. 1977. "Prescribed Burning as a Management Tool," *Proceedings, Symposium on the Environmental Consequences of Fire and Fuel Management in Mediterranean Ecosystems.* U.S. Forest Service, General Technical Report WO-3.

Bjorkbom, J. C. and R. G. Larson. 1977. *The Tionesta Scenic and Research Natural Areas.* Northeastern Forest Experiment Station, General Technical Report NE-31.

Blackburn, W. and P. Tueller. 1970. "Pinyon and Juniper Invasion in Black Sagebrush Communities in East-Central Nevada," *Ecology* 51:841-848.

Bloom, A. 1978. 'Sitka Black-tailed Deer Winter Range in the Kadashan Bay Area, Southeast Alaska," *Journal of Wildlife Management* 42:108-112.

Blydenstein, J. et al. 1957. "Effect of Domestic Livestock Exclusion on Vegetation in the Sonoran Desert," *Ecology* 38:522-526.

Boche, K. E. 1974. *Factors Affecting Meadow-Forest Borders in Yosemite National Park, California,* unpublished M.A. thesis in geography, University of California, Los Angeles.

Bock, J. H. et al. 1978. "A Comparison of Planting and Natural Succession After a Forest Fire in the Northern Sierra Nevada," *Journal of Applied Ecology* 15:597-602.

Bockheim, J. G. et al. No Date. "Impact of Forest Management Practices on Water Quality in Wisconsin" (mimeo).

Boe, K. 1975. *Natural Seedlings and Sprouts After Regeneration Cuttings in Old-Growth Redwood.* Pacific Southwest Forest and Range Experiment Station, Research Paper PSW-111.

Boldt, C. E. et al. 1978. "Riparian Woodlands in Jeopardy on Northern High Plains," pp. 184-189 in *National Symposium on Strategies for Protection and Management of Floodplain Wetlands and Other Riparian Ecosystems.* U.S. Forest Service, General Technical Report WO-12.

Bones, J. T. 1978. *The Forest Resources of West Virginia.* Northeastern Forest Experiment Station, Resource Bulletin NE-56.

Bonnicksen, T. M. and R. G. Lee. 1979. "Persistence of a Fire Exclusion Policy in Southern California: A Biosocial Interpretation," *Journal of Environmental Management* 8:277-293.

Bonnicksen, T. M. and E. Stone. 1981. "The Giant Sequoia — Mixed Conifer Forest Community Characterized Through Pattern Analysis as a Mosaic of Aggregations," *Forest Ecology and Management* 3:307-328.

Boorman, L. A. and R. M. Fuller. 1977. "Studies on the Impact of Paths on the Dune Vegetation at Winterton, Norfolk, England," *Biological Conservation* 12:203-216.

Borchert, J. R. 1950. "The Climate of the Central North American Grassland," *Annals, Association of American Geographers* 40:1-39.

Bormann, F. H. and G. E. Likens. 1979. *Pattern and Process in a Forested Ecosystem.* New York: Springer-Verlag.

Bourdo, E. 1956. "A Review of the General Land Office Survey and of Its Use in Quantitative Studies of Former Forests," *Ecology* 37:754-768.

Bragg, T. and L. Hulbert. 1976. "Woody Plant Invasion of Unburned Kansas Bluestem Prairie," *Journal of Range Management* 29:19-24.

Brander, R. 1974. "Ecological Impacts of Off-Road Vehicles, pp. 29-35 in *Outdoor Recreation Research: Applying the Results.* Northeastern Forest Experiment Station, General Technical Report NC-9.

Brandt, C. J. and R. W. Rhoades. 1973. "Effects of Limestone Dust Accumulation on Lateral Growth of Forest Trees," *Environmental Pollution* 4:207-213.

Branson, F. A. et al. 1972. *Rangeland Hydrology.* Denver: Society for Range Management.

Branson, R. and R. Miller. 1981. "Effects of Increased Precipitation and Grazing Management on Northeastern Montana Rangelands," *Journal of Range Management* 34:3-10.

Bratton, S.P. 1975. "The Effect of the European Wild Boar on Gray Beech Forest in the Great Smoky Mountains," *Ecology* 56:1356-1366.

Bratton, S. P. et al. 1979. "Trail Erosion Patterns in Great Smoky Mountains National Park," *Environmental Management* 3:431-445.

Bratton, S. P. et al. 1980. "Agricultural Area Impacts Within a Natural Area: Cades Cove: A Case Study," *Enviromental Management* 4:433-448.

Bray, J. R. 1957. "Climax Forest Herbs in Prairie," *American Midland Naturalist* 58:434-440.

Brown, R. and E. Farmer. 1976. "Rehabilitation of Alpine Disturbances: Beartooth Plateau, Montana," pp. 58-73 in R. H. Zuck and L. F. Brown (editors), *High-altitude Revegetation Workshop No. 2.* Fort Collins: Colorado State University.

Brown, R. and R. Johnston. 1978a. "Rehabilitation of Disturbed Alpine Rangelands," pp. 704-706 in *Proceedings, First International Rangeland Congress.*

Brown, R. and R. Johnston, 1978b. "Rehabilitation of a High Elevation Mine Disturbance," pp. 116-130 in S. T. Kenny (editor), *High-Altitude Revegetation Workshop No. 3.* Fort Collins: Colorado State University.

Buech, R. et al. 1977. *Small Mammal Populations After a Wildfire in Northeast Minnesota.* North Central Forest Experiment Station, Research Paper NC-151.

Burcham, L. T. 1957. *California Range Land.* Sacramento: California Division of Forestry.

Burcham, L. T. 1970. "Ecological Significance of Alien Plants in California Grasslands," *Proceedings,* Association of American Geographers 2:36-39.

Burkhardt, J. W. and E. W. Tisdale. 1976. "Causes of Juniper Invasion in Southwestern Idaho," *Ecology* 57:472-484.

Burton, T. et al. 1978. "An Approach to the Classification of Utah Mine Spoils and Tailings Based on Surface Hydrology and Erosion," *Environmental Geology* 2:269-278.

Bury, B. et al. 1977. *Effects of Off-Road Vehicles on Vertebrates in the California Desert.* U.S. Fish and Wildlife Service, Research Report 8.

Busack, S. and R. B. Bury. 1974. "Some Effects of Off-Road Vehicles and Sheep Grazing on Lizard Populations in the Mojave Desert," *Biological Conservation* 6:179-183.

California Natural Areas Coordinating Council. No Date. *Inventory of California Natural Areas.* Berkeley; California Natural Areas Coordinating Council.

Cattelino, P. J. et al. 1979. "Predicting the Multiple Pathways of Plant Succession," *Environmental Management* 3:41-50.

Chapin, F. S. and M. C. Chapin. 1980. "Revegetation of an Arctic Disturbed Site by Native Tundra Species," *Journal of Applied Ecology* 17:449-456.

Christensen, E. M. and J. D. Brotherson. 1979. "Decreases of Juniper Woodland in the Utah and Salt Lake Valleys Since Settlement," *Great Basin Naturalist* 39:263-266.

Clawson, M. 1979. "Forests in the Long Sweep of American History," *Science* 204:1168-1174.

Clements, F. E. 1916. *Plant Succession: An Analysis of Development of Vegetation.* Washington, DC: Carnegie Institution Publication 242.

Coates, R. N. and T. O. Miller. 1981. "Cumulative Silvicultural Impacts on Watersheds: A Hydrologic and Regulatory Dilemma," *Environmental Management* 5:147-160.

Cobb, F. and R. Stark. 1970. "Decline and Mortality of Smog-Injured Ponderosa Pine," *Journal of Forestry* 68:147-149.

Coblentz, B. E. 1977. "Some Range Relationships of Feral Goats on Santa Catalina Island, California," *Journal of Range Management* 30:415-419.

Coblentz, B. E. 1978. "The Effects of Feral Goats on Island Ecosystems," *Biological Conservation* 13:279-286.

Cogbill, C. V. 1976. "The Effect of Acid Precipitation on Tree Growth in Eastern North America," pp. 1027-1032 in *First International Symposium on Acid Precipitation and the Forest Ecosystem.* Northeastern Forest Experiment Station, General Technical Report NE-23.

Cole, D. 1977. "Ecosystem Dynamics in the Coniferous Forest of the Willamette Valley, Oregon, U.S.A.," *Journal of Biogeography* 4:181-192.

Cole, D. 1978. "Estimating the Susceptibility of Wildland Vegetation to Alteration," *Journal of Applied Ecology* 15:281-286.

Cole, D. 1979. "Reducing the Impact of Hikers on Vegetation: An Application of Analytical Research Methods," pp. 71-78 in R. Ittner et al. (editors), *Proceedings, Recreational Impacts on Wildlands Conference*, U.S. Forest Service, Pacific Northwest Region.

Cole, D. 1981. "Vegetational Changes Associated With Recreational Use and Fire Suppression in the Eagle Cap Wilderness, Oregon: Some Management Implications," *Biological Conservation* 20:247-270.

Cole, D. 1982. *Wilderness Campsite Impacts: Effect of Amount of Use.* Intermountain Forest and Range Experiment Station, Research Paper INT-284.

Cole, D. and G. Schreiner, 1981. *Impacts of Backcountry Recreation: Site Management and Rehabilitation – An Annotated Bibliography.* Intermountain Forest and Range Experiment Station, General Technical Report INT-121.

Colinvaux, P. A. 1973. *Introduction to Ecology.* New York: John Wiley and Sons.

Connor, R. and C. Adkisson. 1975. "Effects of Clearcutting on the Diversity of Breeding Birds," *Journal of Forestry* 73:781-785.

Conroy, M. *et al.* 1979. "Habitat Components of Clear-Cut Areas for Snowshoe Hares in Michigan," *Journal of Wildlife Management* 43:680-690.

Cooper, C. 1960. "Southwestern Pine Forests Since White Settlement," *Ecological Monographs* 30:129-164.

Corbett, E. S. *et al.* 1978. "Timber Harvesting Practices and Water Quality in the Eastern United States," *Journal of Forestry* 76:484-488.

Cottam, G. 1949. "The Phytosociology of an Oak Woods in Southwestern Wisconsin," *Ecology* 30:271-287.

Cottam, G. and J. Curtis. 1949. "A Method For Making Rapid Surveys of Woodlands by Means of Pairs of Randomly Selected Trees," *Ecology* 30:101-104.

Crouch, G. 1974. "Interaction of Deer and Forest Succession on Clearcuttings in the Coast Range of Oregon," pp. 133-138 in H. Black (editor), *Wildlife and Forest Management in the Pacific Northwest.* Corvallis: Oregon State University.

Crouch, G. 1979. *Long-Term Changes in Cottonwoods on a Grazed and an Ungrazed Plains Bottomland in Northeastern Colorado.* Rocky Mountain Forest and Range Experiment Station, Research Note RM-370.

Cunningham, J. B. *et al.* 1980. *Selection and Use of Snags by Secondary Cavity-Nesting Birds of the Ponderosa Pine Forest,* Rocky Mountain Forest and Range Experiment Station, Research Paper RM-222.

Curtis, J. 1959. *The Vegetation of Wisconsin.* Madison: University of Wisconsin Press.

Curtis, W. R. 1977. *Surface Mining and the Flood of April 1977.* Northeast Forest Experiment Station, Research Note NE-248.

Czapowskyi, M. 1976. *Annotated Bibliography on the Ecology and Reclamation of Drastically Disturbed Areas.* Northeast Forest Experiment Station, General Technical Repot NE-21.

Dale, D. and T. Weaver. 1974. "Trampling Effects on Vegetation of the Trail Corridors of North Rocky Mountain Forests," *Journal of Applied Ecology* 11:767-772.

Dasmann, R. 1964. *Wildlife Biology.* New York: John Wiley and Sons.

Daubenmire, R. 1968. *Plant Communities.* New York: Harper and Row.

Davidson, E. and M. Fox. 1974. "Effects of Off-Road Motorcycle Activity on Mojave Desert Vegetation and Soil," *Madrono* 22:381-390.

Davis, J. B. 1979. "A New Fire Management Policy on Forest Service Lands," *Fire Technology* 15,1:43-50.

Davis, K. M. *et al.* 1980. *Fire Ecology of Lolo National Forest Habitat Types.* International Forest and Range Experiment Station, General Technical Report INT-79.

Davis, M. 1973. "Pollen Evidence of Changing Land Use Around the Shores of Lake Washington," *Northwest Science* 47:133-148.

Davis, M. 1976. "Pleistocene Biogeography of Temperate Deciduous Forest," *Geo-Science and Man* 13:13-26.

Davis, M. 1978. "Climatic Interpretation of Pollen in Quaternary Sediments," pp. 35-51 in D. Walker and J. Guppy (editors), *Biology and Quaternary Environments.* Canberra: Australian Academy of Science.

Davis, O. *et al.* 1977. "Pollen Analysis of Wildcat Lake, Whitman County, Washington: The Last 1000 Years," *Northwest Science* 51:13-30.

Davis, P. 1977. "Cervid Response to Forest Fire and Clearcutting in Southeastern Wyoming," *Journal of Wildlife Management* 41:785-788.

Day, A. D. and K. L. Ludeke. 1980. "Reclamation of Copper Mine Wastes with Shrubs in the Southwestern U.S.A.," *Journal of Arid Environments* 3:107-112.

Day, R. J. 1972. "Stand Structure, Succession, and Use of Southern Albertas Rocky Mountain Forest," *Ecology* 53:472-478.

DeBano, L. *et al.* 1979. "Effects of Burning on Chaparral Soils; I. Soil Nitrogen," *Soil Science of America Journal* 43:504-509.

DeBano, L. 1981. *Water Repellent Soils: A State-of-the-Art.* Pacific Southwest Forest and Range Experiment Station, General Technical Report PSW-46.

DeBenedetti, S. and D. Parsons. 1979. "Natural Fire in Subalpine Meadows: A Case Description from the Sierra Nevada," *Journal of Forestry* 77:477-479.

DeByle, N. 1976. "Soil Fertility as Affected by Broadcast Burning Following Clearcutting in Northern Rocky Mountain Larch/Fir Forests," *Proceedings,* Tall Timbers Fire Ecology Conference 14:447-464.

Delcourt, H. 1975. *Reconstructing the Forest Primeval, West Feliciana Parish.* Louisiana State University, Museum of Geoscience, *Melanges* No. 10.

Delcourt, H. and P. Delcourt. 1974. "Primeval Magnolia-Holly-Beech Climax in Louisiana," *Ecology* 55:638-644.

Delcourt, H. and W. F. Harris. 1980. "Carbon Budget of the Southeastern U.S. Biota: Analysis of Historical Change in Trend from Source to Sink," *Science* 210:321-322.

Delcourt, P. and H. Delcourt. 1979. "Late Pleistocene and Holocene Distributional History of the Deciduous Forest in the Southeastern United States," *Veroff. Geobot. Inst. ETH, Stiftung Rubel, Zurich* 68:79-107.

Delcourt, P. *et al.* 1980. "Quaternary Vegetation History of the Mississippi Embayment," *Quaternary Research* 13:111-132.

Deschamp, J. *et al.* 1979. "Summer Diets of Mule Deer from Lodgepole Pino Habitats," *Journal of Wildlife Management* 43:549-555.

Dickman, A. 1978. "Reduced Fire Frequency Changes Species Composition of a Ponderosa Pine Stand," *Journal of Forestry* 76:24-25.

Dickson, J. and C. Segelquist. 1979. "Breeding Bird Populations in Pine and Pine-Hardwood Forests in Texas," *Journal of Wildlife Management* 43:154-161.

Dodge, J. M. 1975. *Vegetational Changes Associated with Land Use and Fire History in San Diego County,* Unpublished Ph.D. dissertation in geography, University of California, Riverside.

Drury, W. and T. Nisbet. 1973. "Succession," *Arnold Arboretum Journal* 54:331-368.

Dunne, T. and L. Leopold. 1978. *Water in Environmental Planning.* San Francisco: W. H. Freeman.

Dyrness, C. T. 1976. *Effect of Wildfire on Soil Wettability in the High Cascades of Oregon.* Pacific Northwest Forest and Range Experiment Station, Research Paper PNW-202.

Eckert, R. E. *et al.* 1979. "Impacts of Off-Road Vehicles on Infiltration and Sediment Production of Two Desert Soils," *Journal of Range Management* 32:394-397.

Eckstein, R. G. *et al.* 1979. "Snowmobile Effects on Movements of White-tailed Deer: A Case Study," *Environmental Conservation* 6:45-51.

Egler, F. E. 1954. "Vegetation Science Concepts. I. Initial Floristic Composition, A Factor in Old-Field Vegetation Development," *Vegetatio* 14:412-417.

Egler, F. and S. Foote. 1975. *The Plight of the Right of Way Domain: Victim of Vandalism.* Mt. Kisco, New York: Futura Media Services.

Elton, C. 1958. *The Ecology of Invasions by Plants and Animals.* London: Methuen.

England, R. E. and A. DeVos. 1969. "Influence of Animals on Pristine Conditions on the Canadian Grasslands," *Journal of Range Management* 22:87-94.

Farmer, E. et al. 1974. *Revegetation Research on the Decker Coal Mine in Southeastern Montana.* Intermountain Forest and Range Experiment Station, Research Paper INT-162.

Federal Committee on Research Natural Areas. 1968. *A Directory of Research Natural Areas on Federal Lands of the United States of America.* Washington: Superintendent of Documents.

Ferris, C. 1979. "Effects of Interstate 95 on Breeding Birds in Northern Maine," *Journal of Wildlife Management* 43:421-427.

Ffolliott, P. et al. 1977. "Animal Use of Ponderosa Pine Forest Openings," *Journal of Wildlife Management* 41:782-784.

Finley, R. W. 1976. *Original Vegetation of Wisconsin.* St. Paul: North Central Forest Experiment Station.

Fisher, J. B. 1975. "Environmental Impact of Lethal Yellowing Disease of Coconut Palms," *Environmental Conservation* 2:299-304.

Flower, N. 1980. "The Management History and Structure of Unenclosed Woods in the New Forest, Hampshire," *Journal of Biogeography* 7:311-328.

Foin, T. C. et al. 1977. "Quantitative Studies of Visitor Impacts on Environments of Yosemite National Park, California, and Their Implications for Park Management Policy," *Journal of Environmental Management* 5:1-22.

Forcier, L. K. 1975. "Reproductive Strategies and the Co-Occurrence of Climax Tree Species," *Science* 189:808-810.

Franklin, J. 1977. "The Biosphere Reserve Program in the United States," *Science* 195:262-267.

Franklin, J. F. et al. 1972. *Federal Research Natural Areas in Oregon and Washington: A Guidebook for Scientists and Educators.* Portland: Pacific Northwest Forest and Range Experiment Station.

Franzreb, K. E. and R. D. Ohmart. 1978. "The Effects of Timber Harvesting on Breeding Birds in a Mixed-Coniferous Forest," *Condor* 80:431-441.

Frenkel, R. 1974. "Floristic Changes Along Everitt Memorial Highway, Mount Shasta, California," *Wasmann Journal of Biology* 32:105-136.

Gary, H. L. and P. O. Currie. 1977. *The Front Range Pine Type: A 40-Year Photographic Record of Plant Recovery on an Abused Watershed.* Rocky Mountain Forest and Range Experiment Station, General Technical Report RM-46.

Gashwiler, J. S. 1970. "Plant and Mammal Changes on a Clearcut in West-Central Oregon," *Ecology* 51:1018-1026.

Geier, A. R. and L. B. Best. 1980. "Habitat Selection By Small Mammals of Riparian Communities; Evaluating Effects of Habitat Alterations," *Journal of Wildlife Management* 44:16-24.

Gentry, A. H. and J. Lopez-Parodi. 1980. "Deforestation and Increased Flooding of the Upper Amazon," *Science* 210: 1354-1356.

Germain, C. E. 1977. *Wisconsin Scientific Areas.* Wisconsin Department of Natural Resources, Technical Bulletin No. 102.

Gibbens, R. P. and H. F. Heady. 1964. *The Influence of Modern Man on the Vegetation of Yosemite Valley.* California Agricultural Experiment Station, Manual 36.

Goff, F. G. 1967. "Upland Vegetation," pp. 60-89 in G. F. Hanson and F. D. Hole (editors), *Soil Resources and Forest Ecology of Menominee County, Wisconsin.* Madison: University of Wisconsin Extension, Soil Survey Division Bulletin 85.

Goff, F. G. and D. West. 1975. "Canopy-Understory Interaction Effects on Forest Population Structure," *Forest Science* 21:98-108.

Gordon, A. and E. Gorham. 1963. "Ecological Aspects of Air Pollution from an Iron-Sintering Plant at Wawa, Ontario," *Canadian Journal of Botany* 41:1063-1078.

Gorham, E. 1976. "Acid Precipitation and Its Influence Upon Aquatic Ecosystems: An Overview," *Water, Air, and Soil Pollution* 6:457-481.

Gorham, E. and A. Gordon. 1960. "Some Effects of Smelter Pollution Northeast of Falconbridge, Ontario," *Canadian Journal of Botany* 38:307-312.

Gottmann, J. 1961. *Megalopolis.* New York: Twentieth Century Fund.

Gray, D. H. and W. F. Megahan. 1981. *Forest Vegetation Removal and Slope Stability in the Idaho Batholith.* Intermountain Forest and Range Experiment Station, Research Paper INT-271.

Green, L. 1977. *Fuelbreaks and Other Fuel Modification for Wildland Fire Control.* U.S. Forest Service, Agriculture Handbook 499.

Green, L. R. 1981. *Burning By Prescription in Chaparral.* Pacific Southwest Forest and Range Experiment Station, General Technical Report PSW-51.

Greller, A. et al. 1974. "Snowmobile Impact on Three Alpine Tundra Plant Communities," *Environmental Conservation* 1:101-110.

Gross, F. A. and W. A. Dick-Peddie. 1979. "A Map of Primeval Vegetation in New Mexico," *Southwestern Naturalist* 24:115-122.

Gruell, G. 1979. "Wildlife Habitat Investigations and Management Implications on the Bridger-Teton National Forest," pp. 63-73 in M. Bouce and L. Hayden-Wing (editors), *North American Elk, Ecology, Behavior, and Management.* Laramie: University of Wyoming.

Gruell, G. 1980a. *Fire's Influence on Wildlife Habitat on the Bridger-Teton National Forest, Wyoming. Volume 1 – Photographic Record and Analysis.* Intermountain Forest and Range Experiment Station, Research Paper 235.

Gruell, G. 1980b. *Fire's Influence on Wildlife Habitat on the Bridger-Teton National Forest, Wyoming. Volume II – Changes and Causes, Management Implications.* Intermountain Forest and Range Experiment Station, Research Paper INT-252.

Gruell, G. E. and L. L. Loope. 1974. *Relationships Among Aspen, Fire, and Ungulate Browsing in Jackson Hole, Wyoming.* Ogden, UT: U.S. Forest Service and National Park Service.

Habeck, J. and R. Mutch. 1973. "Fire Dependent Forests in the Northern Rocky Mountains," *Quaternary Research* 3:408-424.

Hall, F. C. 1980. "Fire History — Blue Mountains, Oregon," pp. 75-81, in M. A. Stokes and J. H. Dieterich (editors), *Proceedings of the Fire Ecology Workshop.* Rocky Mountain Forest and Range Experiment Station, General Technical Report RM-81.

Halls, L. K. and W. B. Homesley. 1966. "Stand Composition in a Mature Pine-Hardwood Forest of Southeastern Texas," *Journal of Forestry* 64:170-174.

Hamilton, R. and R. Noble. 1975. "Plant Succession and Interactions with Fauna," pp. 96-114 in D. R. Smith (editor), *Proceedings of the Symposium on Management of Forest and Range Habitats for Non-Game Birds.* U.S. Forest Service, General Technical Report WO-1.

Hanley, D. P. et al. 1975. *Stand Structure and Successional Status of Two Spruce-Fir Forests in Southern Utah.* Intermountain Forest and Range Experiment Station, Research Paper INT-176.

Hanley, T. A. and W. W. Brady. 1977. "Feral Burro Impact on a Sonoran Desert Range," *Journal of Range Management* 30:374-377.

Halford, D. K. 1981. "Repopulation and Food Habits of *Peromyscus maniculatus* on a Burned Sagebrush Desert in Southeastern Idaho," *Northwest Science* 55:44-49.

Harniss, R. O. and R. Murray. 1973. "30 Years of Vegetal Change Following Burning of Sagebrush-Grass Range," *Journal of Range Management* 26:322-325.

Harr, R. D. 1976. *Forest Practices and Streamflow in Western Oregon.* Pacific Northwest Forest and Range Experiment Station, General Technical Report PNW-49.

Harr, R. D. 1980. *Streamflow After Patch Logging in Small Drainages Within the Bull Run Municipal Watershed, Oregon.* Pacific Northwest Forest and Range Experiment Station, Research Paper PNW-268.

Harr, R. D. et al. 1975. "Changes in Storm Hydorgraphs After Road Building and Clearcutting in the Oregon Coast Range," *Water Resources Research* 11:436-444.

Harris, D. 1966. "Recent Plant Invasions in the Arid and Semi-Arid Southwest of the United States," *Annals,* Association of American Geographers 56:408-422.

Hart, J. F. 1977. "Land Rotation in Appalachia," *Geographical Review* 67:148-166.

Hastings, J. and R. Turner. 1965. *The Changing Mile.* Tucson: University of Arizona Press.

Hayes, E. M. and J. M. Skelly. 1977. "Transport of Ozone From the Northeast U.S. and Its Effect on Eastern White Pines," *Plant Disease* 61:778-782.

Heady, H. 1975. *Rangeland Management.* New York: McGraw Hill.

Hedgcock, G. G. 1914. "Injuries By Smelter Smoke in Southeastern Tennessee," *Journal,* Washington Academy of Science 4:70-71.

Heinselman, M. L. 1978. "Fire Intensity and Frequency As Factors in the Distribution and Structure of Northern Ecosystems," (Unpublished manuscript).

Helgath, S. 1975. *Trail Deterioration in the Selway-Bitterroot Wilderness.* Intermountain Forest and Range Experiment Station, Research Note INT-193.

Helvey, J. D. et al. 1976. "Some Climatic and Hydrologic Effects of Wildfire in Washington State," *Proceedings,* Tall Timber Fire Ecology Conference 15:201-222.

Henry, J. D. and J. M. A. Swan. 1974. "Reconstructing Forest History from Live and Dead Plant Material — An Approach to the Study of Forest Succession in Southwest New Hampshire," *Ecology* 55:772-783.

Herrington, R. B. and W. G. Beardsley. 1970. Improvement and Maintenance of Campground Vegetation in Central Idaho. Intermountain Forest and Range Experiment Station, Research Paper INT-87.

Hett, J. M. and O. L. Loucks. 1976. "Age Structure of Balsam Fir and Eastern Hemlock," *Journal of Ecology* 64:1029-1044.

Hibbert, A. R. 1979. *"Managing Vegetation to Increase Flow in the Colorado River Basin.* Rocky Mountain Forest and Range Experiment Station, General Technical Report RM-66.

Hill, A. 1976. "The Effects of Man-Induced Erosion and Sedimentation on the Soils of a Portion of the Oak Ridges Moraine," *Canadian Geographer* 20:384-404.

Holmes, D. 1976. *The Effects of Human Trampling and Urine on Subalpine Vegetation: A Survey of Past and Present Backcountry Use, and the Ecological Carrying Capacity of Wilderness, Yosemite National Park,* unpublished M.A. thesis in geography, University of California, Berkeley.

Holzner, W. 1978. "Weed Species and Weed Communities," *Vegetatio* 38:13-20.

Horn, H. S. 1974. "The Ecology of Secondary Succession," *Annual Review of Ecology and Systematics* 5:25-37.

Hosier, P. E. and T. E. Eaton. 1980. "The Impact of Vehicles on Dune and Grassland Vegetation on a South-eastern North Carolina Barrier Beach," *Journal of Applied Ecology* 17:173-182.

Houston, D. B. 1973. "Wildfires in Northern Yellowstone National Park," *Ecology* 54:1111-1117.

Howe, R. and G. Jones. 1977. "Avian Utilization of Small Woodlots in Dane County, Wisconsin," *Passenger Pigeon* 39:313-319.

Humphrey, R. R. and L. A. Mehrhoff. 1958. "Vegetation Changes on a Southern Arizona Grassland Range," *Ecology* 39:720-726.

Iverson, R. M. *et al.* **1981.** "Physical Effects of Vehicular Disturbances on Arid Landscapes," *Science* 212:915-917.

Jarvis, P. J. 1979. "The Ecology of Plant and Animal Introductions," *Progress in Physical Geography* 3:187-214.

Jaynes, R. A. 1978. *A Hydrologic Model of Aspen-Conifer Succession in the Western United States.* Intermountain Forest and Range Experiment Station, Research Paper INT-213.

Jaynes, R. A. and K. Harper. 1978. "Patterns of Natural Revegetation in Arid Southeastern Utah," *Journal of Range Management* 31:407-411.

Jenny, H. *et al.* **1969.** "The Pygmy Forest Podzol Ecosystem and Its Dune Associates of the Mendocino Coast," *Madrono* 20:60-74.

Johannessen, C. *et al.* **1971.** "The Vegetation of the Willamette Valley," *Annals, Association of American Geographers* 61:286-302.

Johnson, F. L. and D. T. Bell. 1975. "Size-Class Structure of Three Streamside Forests," *American Journal of Botany* 62:81-85.

Johnson, M. G. and R. L. Beschta. 1980. "Logging, Infiltration Capacity, and Surface Erodibility in Western Oregon," *Journal of Forestry* 78:334-337.

Johnson, W. C. *et al.* **1979.** "Diversity of Small Mammals in a Powerline Right-of-Way and Adjacent Forest in East Tennessee," *American Midland Naturalist* 101:231-235.

Johnston, A. *et al.* **1971.** "Long-term Grazing Effects on Fescue Grassland Soils," *Journal of Range Management* 24:185-188.

Jones, R. L. and A. H. Beavers. 1964. "Variation of Opal Phytolith Content Among Some Great Soil Groups of Illinois," *Proceedings, Soil Science Society of America* 28:711-712.

Jordan, M. 1975. "Effects of Zinc Smelter Emissions and Fire on a Chestnut-Oak Woodland," *Ecology* 56:78-91.

Jordan, W. R. (editor). 1980. *A Guide to the University of Wisconsin-Madison Arboretum: Gallistel Woods.* Madison: Friends of the University of Wisconsin-Madison Arboretum.

Jurgensen, M. F. *et al.* **1979.** *Forest Soil Biology – Timber Harvesting Relationships.* Intermountain Forest and Range Experiment Station, General Technical Report INT-69.

Kapp, R. 1977. "Late Pleistocene and Postglacial Plant Communities of the Great Lakes Region," pp. 1-26 in R. C. Romans (editor), *Geobotany.* New York: Plenum Press.

Karnosky, D. F. 1979. "Dutch Elm Disease: A Review of the History, Environmental Implications, Control, and Research Needs," *Environmental Conservation* 6:311-322.

Karr, J. 1968. "Habitat and Avian Diversity on Strip-Mined Land in East-Central Illinois," *Condor* 70:348-357.

Keay, J. and J. Peek. 1980. "Relationships Between Fires and Winter Habitat of Deer in Idaho," *Journal of Wildlife Management* 44:372-380.

Kelty, M. and R. Nyland. 1981. "Regenerating Adirondak Northern Hardwoods by Shelterwood Cutting and Control of Deer Density," *Journal of Forestry* 79:22-26.

Kilgore, B. 1973. "Impact of Prescribed Burning on a Sequoia-Mixed Conifer Forest," *Proceedings,* Tall Timbers Fire Ecology Conference 12:345-375.

Kilgore, B. 1976. "Fire Management in the National Parks: An Overview," *Proceedings,* Tall Timbers Fire Ecology Conference 14:45-57.

Kilgore, B. and D. Taylor. 1979. "Fire History of a Sequoia-Mixed Conifer Forest," *Ecology* 60:129-142.

Kimmins, J. P. 1977. "Evaluation of the Consequences for Future Tree Productivity of the Loss of Nutrients in Whole-Tree Harvesting," *Forest Ecology and Management* 1:169-183.

Kirkland, G. L. 1977. "Responses of Small Mammals to the Clearcutting of Northern Appalachian Forests," *Journal of Mammalogy* 58:600-609.

Knox, J. C. 1977. "Human Impacts on Wisconsin Stream Channels," *Annals,* Association of American Geographers 67:323-342.

Koehler, G. et al. 1975. "Preserving the Pine Marten: Management Guidelines for Western Forests," *Western Wildlands* 2,3:31-36.

Koehler, G. and M. Hornocker. 1977. "Fire Effects on Marten Habitat in the Selway-Bitterroot Wilderness," *Journal of Wildlife Management* 41:500-505.

Koford, C. 1958. *Prairie Dogs, Whitefaces, and Blue Grama.* Wildlife Society, Wildlife Monograph No. 3.

Komarek, E. V. 1974. "Effects of Fire on Temperate Forests and Related Ecosystems: Southeastern United States," pp. 251-277 in Kozlowski and Ahlgren (1974).

Kozlowski, T. T. and C. E. Ahlgren (editors). 1974. *Fire and Ecosystems.* New York: Academic Press.

Krebill, R. G. 1972. *Mortality of Aspen on the Gros Ventre Elk Winter Range.* Intermountain Forest and Range Experiment Station, Research Paper INT-129.

Krefting, L. and C. Ahlgren. 1974. "Small Mammals and Vegetation Changes After Fire in a Mixed Conifer-Hardwood Forest," *Ecology* 55:1391-1398.

Kruse, W. H. et al. 1979. "Community Development in Two Adjacent Pinyon-Juniper Eradication Areas Twenty-five Years After Treatment," *Journal of Environmental Management* 8:237-247.

LaMarche, V. C. and H. A. Mooney. 1972. "Recent Climatic Change and Development of the Bristlecone Pine Krummholz Zone, Mt. Washington, Nevada," *Arctic and Alpine Research* 4:61-72.

LaPage, W. 1967. *Some Observations on Campground Trampling and Ground Cover Response.* Northeastern Forest Experiment Station, Research Paper NE-68.

Lathrop, E. W. and E. F. Archbold. 1980a. "Plant Response to Los Angeles Aqueduct Construction in the Mojave Desert," *Environmental Management* 4:137-148.

Lathrop, E. W. and E. F. Archbold. 1980b. "Plant Response to Utility Right of Way Construction in the Mojave Desert," *Environmental Management* 4:215-226.

Laycock, W. A. 1969. *Exclosures and Natural Areas on Rangelands in Utah.* Intermountain Forest and Range Experiment Station, Research Paper INT-62.

Laycock, W. A. and P. W. Conrad. 1981. "Responses of Vegetation and Cattle to Various Systems of Grazing on Seeded and Native Mountain Rangelands in Eastern Utah," *Journal of Range Management* 34:52-58.

Leak, W. 1964. "An Expression of Diameter Distribution for Unbalanced, Uneven-Aged Stands and Forests," *Forest Science* 10:39-50.

Leak, W. 1975. "Age Distribution in Virgin Red Spruce and Northern Hardwoods," *Ecology* 56:1451-1454.

Leak, W. and R. Graber. 1974. "A Method for Detecting Migration of Forest Vegetation," *Ecology* 55:1425-1427.

Legge, A. H. 1980. "Primary Productivity, Sulfur Dioxide, and the Forest Ecosystem: An Overview of a Case Study," pp. 51-62 in Miller (editor 1980).

Leitner, L. A. and M. T. Jackson. 1981. "Presettlement Forest of the Unglaciated Portion of Southern Illinois," *American Midland Naturalist* 105:290-304.

Leopold, A. S. 1966. "Adaptability of Animals to Habitat Change," pp. 66-75 in F. F. Darling and J. P. Milton (editors), *Future Environments of North America.* New York: Natural History Press.

Leopold, A. S. 1959. "Big Game Management," pp. 85-99 in *Survey of Fish and Game Problems in Nevada.* Nevada Legislative Council Bulletin 36.

Leopold, A. S. and F. F. Darling. 1955. *Wildlife in Alaska.* New York: Ronald Press.

Leopold, L. 1951. "Vegetation of Southwestern Watersheds in the Nineteenth Century," *Geographical Review* 41:295-316.

Lewis, J. 1969. "Range Management Viewed in the Ecosystem Framework," pp. 97-187 in G. M. Van Dyne (editor), *The Ecosystem Concept in Natural Resource Management.* New York: Academic Press.

Lewis, W. M. and M. C. Grant. 1980. "Acid Precipitation in the Western United States," *Science* 207:176-177.

Liddle, M. J. 1975. "A Selective Review of the Ecological Effects of Human Trampling on Natural Ecosystems," *Biological Conservation* 7:17-36.

Liddle, M. J. and P. Greig-Smith. 1975. "A Survey of Tracks and Paths in a Sand Dune Ecosystem. I. Soils. II. Vegetation," *Journal of Applied Ecology* 12:893-930.

Liddle, M. J. and H. R. A. Scorgie. 1980. "The Effects of Recreation on Freshwater Plants and Animals: A Review," *Biological Conservation* 17:183-206.

Likens, G. E. and F. H. Bormann. 1974. "Acid Rain: A Serious Regional Environmental Problem," *Science* 184:1176-1179.

Lillywhite, H. B. 1977. "Effects of Chaparral Conversion on Small Vertebrates in Southern California," *Biological Conservation* 11:171-184.

Lima, W. P. et al. 1978. *Natural Reforestation Reclaims a Watershed: A Case History from West Virginia.* Northeastern Forest Experiment Station, Research Paper NE-392.

Linzon, S. N. 1971. "Economic Effects of Sulphur Dioxide on Forest Growth," *Journal, Air Pollution Control Association* 21:81-86.

Loope, L. and G. Gruell. 1973. "The Ecological Role of Fire in the Jackson Hole Area, Northwestern Wyoming," *Quaternary Research* 3:425-443.

Lorimer, C. G. 1977. "The Presettlement Forest and Natural Disturbance Cycle of Northeastern Maine," *Ecology* 58:139-148.

Lorimer, C. G. 1980. "Age Structure and Disturbance History of a Southern Appalachian Virgin Forest," *Ecology* 61:1169-1184.

Loucks, O. 1970. "Evolution of Diversity, Efficiency, and Community Stability," *American Zoologist* 10:17-25.

Lucich, G. and R. Hansen. 1981. "Autumn Mule Deer Foods on Heavily Grazed Cattle Ranges in Northwestern Colorado," *Journal of Range Management* 34:72-73.

Luckenbach, R. 1975. "What the ORV's Are Doing to the Desert," *Fremontia* 2,4:3-11.

Ludwig, J. et al. 1977. "An Evaluation of Transmission Line Construction on Pinon-Juniper Woodland and Grassland Communities in New Mexico," *Journal of Environmental Management* 5:127-137.

Lutz, H. J. 1930. "Original Forest Composition in Northwestern Pennsylvania as Indicated by Early Land Survey Notes," *Journal of Forestry* 28:1098-1103.

Lyon, L. J. and P. Stickney. 1976. "Early Vegetal Succession Following Large Northern Rocky Mountain Wildfires," *Proceedings,* Tall Timber Fire Ecology Conference 14:355-375.

Lyon, L. J. and C. Jensen. 1980. "Management Implications of Elk and Deer Use of Clear-Cuts in Montana," *Journal of Wildlife Management* 44:352-362.

Mackie, R. J. 1978. "Impacts of Livestock Grazing on Wild Ungulates," *Transactions, North American Wildlife and Natural Resources Conference* 43:462-476.

Magill, A. 1970. *Five California Campgrounds: Conditions Improve After 5 Years' Recreational Use.* Pacific Southwest Forest and Range Experiment Station, Research Paper PSW-62.

Marr, J. W. 1977. "The Development and Movement of Tree Islands Near the Upper Limit of Tree Growth in the Southern Rocky Mountains," *Ecology* 58:1159-1164.

Marquis, D. 1974. *The Impact of Deer Browsing on Allegheny Hardwood Regeneration.* Northeastern Forest Experiment Station, Research Paper NE-308.

Marquis, D. 1975. *The Allegheny Hardwood Forests of Pennsylvania.* Northeastern Forest Experiment Station, General Technical Report NE-15.

Marquis, D. and T. J. Grisez. 1978. *The Effect of Deer Exclosures on the Recovering of Vegetation in Failed Clearcuts on the Allegheny Plateau.* Northeast Forest and Range Experiment Station, Research Note NE-270.

Marquiss, R. and R. Land. 1959. "Vegetational Composition and Ground Cover of Two Natural Relict Areas and Their Associated Grazed Areas in the Red Desert of Wyoming," *Journal of Range Management* 12:104-109.

Marsh, G. P. 1864. *Man and Nature.* New York: Charles Scribner and Company.

Marshall, E. 1981. "The Summer of the Gypsy Moth," *Science* 213:991-993.

Martin, R. E. et al. 1976. "Fire in the Pacific Northwest — Perspectives and Problems," *Proceedings,* Annual Tall Timbers Fire Ecology Conference 15:1-23.

Martin, S. C. and R. Turner. 1977. "Vegetation Change in the Sonoran Desert Region, Arizona and Sonora," *Journal of the Arizona Academy of Science* 12:59-69.

McAtee, J. W. and D. L. Drawe. 1981. "Human Impact on Beach and Foredune Vegetation of North Padre Island, Texas," *Environmental Management* 4:527-538.

McBride, J. R. 1974. "Plant Succession in the Berkeley Hills, California," *Madrono* 22:317-329.

McBride, J. R. and D. Jacobs. 1976. "Urban Forest Development: A Case Study. Menlo Park, California," *Urban Ecology* 2:1-14.

McBride, J. R. and R. D. Laven. 1976. "Scars as an Indicator of Fire Frequency in the San Bernardino Mountains, California," *Journal of Forestry* 74:439-442.

McClelland, B. R. et al. 1979. "Habitat Management for Hole-Nesting Birds in Forests of Western Larch and Douglas Fir," *Journal of Forestry* 77:480-483.

McCauley, O. and G. Trimble. 1975. *Site Quality in Appalachian Hardwoods: The Biological and Economic Response Under Selection Silviculture.* Northeastern Forest Experiment Station, Research Paper NE-312.

McCormick, J. F. and R. B. Platt. 1980. "Recovery of an Appalachian Forest Following the Chestnut Blight, or Catherine Keever — You Were Right!" *American Midland Naturalist* 104:264-273.

McDonald, P. M. 1976. *Forest Regeneration and Seedling Growth From Five Major Cutting Methods in North-Central California.* Pacific Southwest Forest and Range Experiment Station, Research Paper PSW-115.

McGinty, W. A. et al. 1978. "Influence of Soil, Vegetation, and Grazing Management on Infiltration Rate and Sediment Production of Edwards Plateau Rangeland," *Journal of Range Management* 32:33-37.

McIntosh, R. P. 1979. "The Relation Between Succession and the Recovery Process in Ecosystems," pp. 11-62 in J. Cairns (editor), *The Recovery Process in Damaged Ecosystems.* Ann Arbor, MI: Ann Arbor Publications.

McNeil, R. C. and D. B. Zobel. 1980. "Vegetation and Fire History of a Ponderosa Pine-White Fir Forest in Crater Lake National Park," *Northwest Science* 54:30-46.

McNicol, J. G. and F. F. Gilbert. 1980. "Late Winter Use of Upland Cutovers by Moose," *Journal of Wildlife Management* 44:363-371.

McKnight, T. 1958. "The Feral Burro in the U.S. : Distributions and Problems," *Journal of Wildlife Management* 22:163-178.

Meeuwig, R. O. and P. E. Packer. 1976. "Erosion and Runoff on Forest and Range Lands," pp. 105-116 in H. F. Heady *et al.* (editors), *Watershed Management on Range and Forest Lands.* Logan: Utah State University.

Megahan, W. F. and D. C. Moliter. 1975. "Erosional Effects of Wildfire and Logging in Idaho," pp. 423-444 in *Proceedings, Watershed Management Symposium,* American Society of Civil Engineers.

Megahan, W. F. and W. Kidd. 1972. *Effect of Logging Roads on Sediment Production Rates in the Idaho Batholith.* Intermountain Forest and Range Experiment Station, Research Paper INT-123.

Metzger, F. 1980. *Strip Clearcutting to Regenerate Northern Hardwoods.* North Central Forest Experiment Station, Research Paper NC-186.

Meyer, H. A. 1952. "Structure, Growth, and Drain in Balanced Uneven-Aged Forests," *Journal of Forestry* 50:85-92.

Miles, S. R. and P. C. Singleton. 1975. "Vegetative History of Cinnabar Park in Medicine Bow National Forest, Wyoming," *Proceedings,* Soil Science Society of America 39:1204-1208.

Miller P. R. (editor). 1980. *Proceedings of Symposium on Effects of Air Pollutants on Mediterranean and Temperate Forest Ecosystems.* Pacific Southwest Forest and Range Experiment Station, General Technical Report PSW-43.

Minore, D. 1978. *The Dead Indian Plateau: A Historical Summary of Forestry Observations and Research in a Severe Southwestern Oregon Environment.* Pacific Northwest Forest and Range Experiment Station, General Technical Report PNW-72.

Moir, W. H. 1972. "Natural Areas," *Science* 177:396-400.

Monsen, S. B. and A. P. Plummer, 1978. "Plants and Treatment for Revegetation of Disturbed Sites in the Intermountain Area," pp. 155-173 in R. A. Wright (editor), *The Reclamation of Disturbed Arid Lands.* Albuquerque: University of New Mexico Press.

Moss, M. 1976. "Forest Regeneration in the Rural-Urban Fringe: A Study of Secondary Succession in the Niagara Peninsula," *Canadian Geographer* 20:141-157.

Mudd, J. B. and T. T. Kozlowski (editors). 1975. *Response of Plants to Air Pollution.* New York: Academic Press.

Mueggler, W. F. and D. L. Bartos. 1977. *Grindstone Flat and Big Flat Exclosures – A 41-Year Record of Changes in Clearcut Aspen Communities.* Intermountain Forest and Range Experiment Station, Research Paper INT-195.

Mueller-Dombois, D. 1972. "Crown Distortion and Elephant Distribution in the Woody Vegetations of Ruhuna National Park, Ceylon," *Ecology* 53:208-226.

Mulcahy, D. L. 1975. "Differential Mortality Among Cohorts in a Population of *Acer saccharum* Seedlings," *American Journal of Botany* 62:422-426.

Muller, P. R. and J. R. McBride. 1975. "Effects of Air Pollutants on Forests," pp. 192-235 in J. B. Mudd and T. T. Kozlowski (editors), *Responses of Plants to Air Pollution.* New York: Academic Press.

Muller, R. A. 1971. "Frequency Analyses of the Ratio of Actual to Potential Evapotranspiration for the Study of Climate and Vegetation Relationships," *Proceedings,* Association of American Geographers 3:118-122.

Mutch, R. W. 1970. "Wildland Fires and Ecosystems — A Hypothesis," *Ecology* 51:1046-1051.

Myers, N. 1972. *The Long African Day.* New York: Macmillan.

Neuenschwander, L. F. 1980. "Broadcast Burning of Sagebrush in the Winter," *Journal of Range Management* 33:233-236.

Neumann, P. and H. Merriam. 1972. "Ecological Effects of Snowmobiles," *Canadian Field Naturalist* 86:207-212.

Niering, W. A. et al. 1963. "The Saguaro: A Population in Relation to Environment," *Science* 142:15-23.

Niering, W. A. and R. Goodwin. 1974. "Creation of Relatively Stable Shrublands with Herbicides: Arresting 'Succession' on Rights-of-Way and Pastureland," *Ecology* 55:784-795.

Odum, E. 1959. *Fundamentals of Ecology.* Philadelphia: W. B. Saunders.

Odum, E. 1969. "The Strategy of Ecosystem Development," *Science* 164:262-270.

Oliver, C. D. 1981. "Forest Development in North America Following Disturbances," *Forest Ecology and Management* 3:153-168.

Oliver, C. D. and E. P. Stephens. 1977. "Reconstruction of a Mixed-Species Forest in Central New England," *Ecology* 58:562-572.

Oliver, W. W. 1979. *Fifteen-Year Growth Patterns After Thinning a Ponderosa-Jeffrey Pine Plantation in Northeastern California.* Pacific Southwest Forest and Range Experiment Station, Research Paper PSW-141.

O'Meara, T. E. et al. 1981. "Nongame Wildlife Responses to Chaining of Pinyon-Juniper Woodlands," *Journal of Wildlife Management* 45:381-389.

Overrein, L. N. 1980. *Acid Precipitation Impacts on Terrestrial and Aquatic Systems in Norway.* Pacific Southwest Forest and Range Experiment Station, General Technical Report PSW-43.

Painter, E. L. and J. K. Detling. 1981. "Effects of Defoliation on Net Photosynthesis and Regrowth of Western Wheatgrass," *Journal of Range Management* 34:68-71.

Parker, A. J. 1980. "The Successional Status of *Cupressus arizonica*," *Great Basin Naturalist* 40:254-264.

Parsons, D. 1976. "The Role of Fire in Natural Communities: An Example from the Southern Sierra Nevada, California," *Environmental Conservation* 3:91-99.

Parsons, D. 1981. "The Historical Role of Fire in the Foothill Communities of Sequoia National Park," *Madrono* 28:111-120.

Parsons, D. and S. DeBenedetti. 1979. "Impact of Fire Suppression on a Mixed-Conifer Forest," *Forest Ecology and Management* 2:21-33.

Patric, J. H. and G. M. Aubertin. 1977. "Long-Term Effects of Repeated Logging on an Appalachian Stream," *Journal of Forestry* 75:492-494.

Patton, D. 1976. *Timber Harvesting Increases Deer and Elk Use of a Mixed Conifer Forest.* Rocky Mountain Forest and Range Experiment Station, Research Note RM-329.

Patton, D. 1977. "Managing Southwestern Ponderosa Pine for the Abert Squirrel," *Journal of Forestry* 75:264-267.

Pechanec, J. F. et al. 1965. *Sagebrush Control on Rangelands.* U.S. Department of Agriculture, Handbook 277.

Peek, J. et al. 1979. "Evaluation of Fall Burning on Bighorn Sheep Winter Range," *Journal of Range Management* 32:430-432.

Phillips, E. 1976. "Changes in the Phytosociology of Boreal Conifer-Hardwood Forests on Mackinac Island, Michigan, 1934-1974," *American Midland Naturalist* 96:317-323.

Phillips, W. S. 1963. *Vegetational Changes in Northern Great Plains.* University of Arizona, Agricultural Experiment Station Report 214.

Pitt, M. and H. Heady. 1979. "The Effects of Grazing Intensity on Annual Vegetation," *Journal of Range Management* 32:109-114.

Pitt, M.D. *et al.* **1978.** "Influences of Brush Conversion and Weather Patterns on Runoff from a Northern California Watershed," *Journal of Range Management* 31:23-27.

Powell, D. S. and N. P. Kingsley. 1980. *The Forest Resources of Maryland.* Northeastern Forest Experiment Station, Resource Bulletin NE-61.

Rankin, H. T. and D. E. Davis. 1971. "Woody Vegetation in the Black Belt Prairie of Montgomery County, Alabama, in 1845-46," *Ecology* 52:716-719.

Ream, C. H. 1981. *The Effects of Fire and Other Disturbances on Small Mammals and Their Predators: An Annotated Bibliography.* Intermountain Forest and Range Experiment Station, General Technical Report INT-106.

Reiners, W. A. and G. E. Lang. 1979. "Vegetational Patterns and Processes in the Balsam Fir Zone, White Mountains, New Hampshire," *Ecology* 60:403-417.

Reynolds, T. 1979. "Response of Reptile Populations to Different Land Management Practices on the Idaho National Engineering Laboratory Site," *Great Basin Naturalist* 39:255-262.

Reynolds, T. and C. Trost. 1981. "Grazing, Crested Wheatgrass, and Bird Populations in Southeastern Idaho," *Northwest Science* 55:225-234.

Rice, R. M. *et al.* **1979.** *A Watershed's Response to Logging and Roads: South Fork of Caspar Creek, California, 1967-1976.* Pacific Southwest Forest and Range Experiment Station, Research Paper PSW 146.

Rickard, W. and J. Brown. 1974. "Effects of Vehicles on Arctic Tundra," *Environmental Conservation* 1:55-62.

Robertson, J. H. 1971. "Changes on a Sagebrush-Grass Range in Nevada Ungrazed for 30 Years," *Journal of Range Management* 24:397-400.

Robertson, J. and C. K. Pearse, 1945. "Artificial Reseeding and the Closed Community," *Northwest Science* 19:58-66.

Robinson, T. W. 1965. *Introduction, Spread and Areal Extent of Saltcedar in the Western States.* U.S. Geological Survey, Professional Paper 491-A.

Roby, G. A. and L. Green. 1976. *Mechanical Methods of Chaparral Modification.* U.S. Forest Service, Agriculture Handbook 487.

Rogers, G. F. 1980. *Photographic Documentation of Vegetation Change in the Great Basin Desert,* unpublished Ph.D. dissertation in geography, University of Utah, Salt Lake City.

Romme, W. H. and D. H. Knight. 1981. "Fire Frequency and Subalpine Forest Succession Along a Topographic Gradient in Wyoming," *Ecology* 62:319-326.

Ross, B. A. *et al.* **1970.** "Effects of Long-Term Deer Exclusion on a *Pinus resinosa* Forest in North-Central Minnesota," *Ecology* 51:1088-1093.

Rowntree, R. A. and J. L. Wolfe. 1980. *Abstracts of Urban Forestry: Research in Progress – 1979.* Northeastern Forest Experiment Station, General Technical Report NE-60.

Rundel, P. W. *et al.* **1977.** "Montane and Subalpine Vegetation of the Sierra Nevada and Cascade Ranges," pp. 559-599 in M. G. Barbour and J. Major (editors), *Terrestrial Vegetation of California.* New York: John Wiley and Sons.

Runnell, R. 1951. "Some Effects of Livestock Grazing on Ponderosa Pine Forest and Range in Central Washington," *Ecology* 32:594-607.

Russell, E. W. B. 1981. "Vegetation of Northern New Jersey Before European Settlement," *American Midland Naturalist* 105:1-12.

Sauer, C. O. 1950. "Grassland Climax, Fire, and Man," *Journal of Range Management* 3:16-21.

Schmelz, D. V. 1975. "Donaldson's Woods: Two Decades of Change," *Ecology* 84:234-243.

Schmid, J. 1975. *Urban Vegetation.* University of Chicago, Department of Geography, Research Paper 161.

Schmid, J. 1979. "Vegetation Types, Functions, and Constraints in Metropolitan Environments," pp. 499-528 in M. T. Beatty *et al.* (editors), *Planning the Uses and Management of Land.* Madison: American Society of Agronomy.

Schier, G. 1975. *Deterioration of Aspen Clones in the Middle Rocky Mountains.* Intermountain Forest and Range Experiment Station, Research Paper INT-170.

Schmutz, E. 1967. "Boysag Point: A Relict Area on the North Rim of Grand Canyon in Arizona," *Journal of Range Management* 20:363-369.

Schmutz, E. and D. Smith. 1976. "Successional Classification of Plants on a Desert Grassland Site in Arizona," *Journal of Range Management* 29:476-479.

Schofield, E. et al. 1970. "Probable Damage to Tundra Biota Through Sulphur Dioxide Destruction of Lichens," *Biological Conservation* 2:278-280.

Scholl, D. G. 1975. "Soil Wettability and Fire in Arizona Chaparral," *Proceedings, Soil Science Society of America* 39:356-361.

Severson, K. and Kranz, J. 1978. "Management of Bur Oak on Deer Winter Range," *Bulletin,* Wildlife Society 6:212-216.

Shanfield, A. 1981. "Alder, Cottonwood, and Sycamore Regeneration and Distribution Along the Nacimiento River, California," paper presented at the California Riparian Systems Conference, University of California at Davis, September 17-19.

Shinn, D. A. 1980. "Historical Perspectives on Range Burning in the Inland Pacific Northwest," *Journal of Range Management* 33:415-423.

Short, H. et al. 1977. "The Use of Natural and Modified Pinyon Pine-Juniper Woodlands by Deer and Elk," *Journal of Wildlife Management* 41:543-559.

Siccama, T. G. 1971. "Presettlement and Present Forest Vegetation in Northern Vermont with Special Reference to Chittenden County," *American Midland Naturalist* 85:153-172.

Simmons, I. and T. Vale. 1975. "Conservation of the California Coast Redwood and Its Environment," *Environmental Conservation* 2:29-38.

Sims, H. P. and C. Buckner. 1973. "The Effect of Clear Cutting and Burning of *Pinus banksiana* Forests on the Populations of Small Mammals in Southestern Manitoba," *American Midland Naturalist* 90:228-231.

Singer, F. J. 1981. "Wild Pig Populations in the National Parks," *Environmental Management* 5:263-270.

Skelly, J. M. 1980. "Photochemical Oxidant Impact on Mediterranean and Temperate Forest Ecosystems: Real and Potential Effects," pp. 38-50 in Miller (1980).

Skelly, J. M. et al. 1979. "Impact of Photochemical Oxidant Air Pollution on Eastern White Pine in The Shenandoah, Blue Ridge Parkway, and Great Smoky Mountains National Parks," in *Proceedings, Second Conference on Scientific Research in the National Parks.* Washington: Government Printing Office.

Sly, G. 1976. "Small Mammal Succession on Strip-Mined Land in Vigo County, Indiana," *American Midland Naturalist* 95:257-267.

Smith, W. 1974. "Air Pollution — Effects on the Structure and Function of the Temperate Forest Ecosystem," *Environmental Pollution* 6:111-129.

Smith, M. et al. 1979. "Forage Selection by Mule Deer on Winter Range Grazed by Sheep in Spring," *Journal of Range Management* 32:40-45.

Smith, P. C. 1980. "California Conifers Thrive in New Zealand," *California Agriculture* 34, 8:4-6.

Smith, W. H. 1980. "Air Pollution — A 20th Century Allogenic Influence on Forest Ecosystems," pp. 79-87 in Miller (1980).

Snyder, J. D. and R. A. Janke. 1976. "Impact of Moose Browsing on Boreal-Type Forests of Isle Royal National Park," *American Midland Naturalist* 95:79-92.

Soutiere, E. 1979. "Effects of Timber Harvesting on Marten in Maine," *Journal of Wildlife Management* 43:850-860.

Spatz, G. and D. Mueller-Dombois. 1973. "The Influence of Feral Goats on Koa Tree Reproduction in Hawaii Volcanoes National Park," *Ecology* 54:870-876.

Sprugel, D. G. 1976. "Dynamic Structure of Wave-Regenerated *Abies balsamea* Forests in the North-Eastern United States," *Journal of Ecology* 64:889-911.

Sprugel, D. G. and F. H. Bormann. 1981. "Natural Disturbance and the Steady State in High-Altitude Balsam Fir Forests," *Science* 211:390-393.

Spurr, S. H. 1979. "Silviculture," *Scientific American* 240, 2:76-91.

Spurr, S. H. and B. V. Barnes. 1980. *Forest Ecology* (Third Edition). New York: John Wiley and Sons.

Stauffer, D. and L. Best. 1980. "Habitat Selection by Birds of Riparian Communities: Evaluating Effects of Habitat Alterations," *Journal of Wildlife Management* 44:1-15.

Stearns, F. 1949. "Ninety Years Change in a Northern Hardwood Forest in Wisconsin," *Ecology* 30:350-358.

Stebbins, R. and N. Cohen. 1976. "Off-Road Menace," *Sierra Club Bulletin* 61, 7:33-37.

Stephens, E. P. 1956. "The Uprooting of Trees: A Forest Process," *Proceedings,* Soil Science Society of America 20:113-116.

Stevens, D. 1966. "Range Relationships of Elk and Livestock, Crow Creek Drainage, Montana," *Journal of Wildlife Management* 30:349-363.

Stewart, G. and A. C. Hull. 1949. "Cheatgrass — An Ecologic Intruder in Southern Idaho," *Ecology* 30:58-74.

Stoddart, L. et al. 1975. *Range Management.* New York: McGraw-Hill.

Strelke, W. and J. Dickson. 1980. "Effect of Forest Clear-Cut Edge on Breeding Birds in East Texas," *Journal of Wildlife Management* 44:559-567.

Stokes, M.A. and J. H. Dieterich. 1980. *Proceedings of the Fire Ecology Workshop.* Rocky Mountain Forest and Range Experiment Station, General Technical Report RM-8.

Stroessner, W. J. and J. R. Habeck. 1966. "The Presettlement Vegetation of Iowa County, Wisconsin," *Transactions,* Wisconsin Academy of Sciences, Arts, and Letters 55:167-180.

Swain, A. M. 1980. "Landscape Patterns and Forest History in the Boundary Waters Canoe Area. Minnesota: A Pollen Study from Hug Lake," *Ecology* 61:747-754.

Swanson, F. and C. Dyrness. 1975. "Impact of Clear-cutting and Road Construction on Soil Erosion by Landslides in the Western Cascade Range, Oregon," *Geology* 3,7:393-396.

Swanston, D. and F. Swanson. 1976. "Timber Harvesting, Mass Erosion, and Steepland Forest Geomorphology in the Pacific Northwest," pp. 199-221 In D. Coates (editor), *Geomorphology and Engineering.* Stroudsburg, Pennsylvania: Hutchinson and Ross.

Szaro, R. and R. Balda. 1979. *Effects of Harvesting Ponderosa Pine on Nongame Bird Populations.* Rocky Mountain Forest and Range Experiment Station, Research Paper RM-212.

Tans, W. 1976. *The Presettlement Vegetation of Columbia County, Wisconsin, in the 1830's.* Wisconsin Deaprtment of Natural Resources, Technical Bulletin 90.

Taylor, D. L. 1973. "Some Ecological Implications of Forest Fire Control in Yellowstone National Park, Wyoming," *Ecology* 54:1394-1396.

Taylor, D. L. 1980. "Fire History and Man-induced Fire Problems in Subtropical South Florida," pp. 63-68 in Stokes and Dieterich (1980).

Thomas, J. W. (editor). 1979. *Wildlife Habitats in Managed Forests: The Blue Mountains of Oregon and Washington.* U.S. Forest Service Agricultural Handbook 553.

Thomas, J. W. et al. 1975. "Silvicultural Options and Habitat Values in Coniferous Forests," pp. 272-287 in D. R. Smith (editor), *Proceedings, Symposium on Management of Forest and Range Habitats for Nongame Birds.* U.S. Forest Service, General Technical Report WO-1.

Thompson, K. 1961. "Riparian Forests of the Sacramento Valley, California," *Annals,* Association of American Geographers 51:294-315.

Titterington, R. W. et al. 1979. "Songbird Responses to Commercial Clear-Cutting in Maine Spruce-Fir Forests," *Journal of Range Management* 43:602-609.

Treshow, M. 1980. "Pollution Effects on Plant Distribution," *Environmental Conservation* 7:279-286.

Trewartha, G. 1940. "The Vegetal Cover of the Driftless Cuestaform Hill Lands: Presettlement Record and Postglacial Evolution," *Transactions,* Wisconsin Academy of Sciences, Arts, and Letters 32:361-382.

Trout, L. and T. Leege. 1971. "Are the Northern Idaho Elk Herds Doomed?" *Idaho Wildlife Review* 24,3:3-6.

Twight, P. 1973. *Ecological Forestry for the Douglas Fir Region.* Washington, DC: National Parks and Conservation Association.

Twight, P. and L. Minckler. 1972. *Ecological Forestry for the Central Hardwood Forest.* Washington, DC: National Parks and Conservation Association.

U.S. Bureau of Land Management. 1979a. *Historical Comparison Photography: Missouri Breaks, Montana.* Billings: Montana State Office.

U.S. Bureau of Land Management. 1979b. *Historical Comparison Photography: Mountain Foothills, Dillon Research Area, Montana.* Billings: Montana State Office.

U.S. Forest Service. 1971. *Regrowing America's Forests: A 30-Year Photographic Record.* Washington, DC. U.S. Forest Service.

U.S. Forest Service. 1973. *Silvicultural Systems for the Major Forest Types of the United States.* Washington, DC: Government Printing Office.

U.S. Senate. 1971. " 'Clearcutting' Practices on National Timberlands," Hearings before the Subcommittee on Public Lands of the Committee on Interior and Insular Affairs, U.S. Senate, 92nd Congress, 1st Session, April 5 and 6, Part 1.

Vale, T.R. 1973. "The U.S. Geological Survey Photographic Library," *Historical Geography Newsletter* 3,2:14.

Vale, T.R. 1974. "Sagebrush Conversion Projects: An Element of Contemporary Environmental Change in the Western United States," *Biological Conservation* 6:274-284.

Vale, T.R. 1975a. "Presettlement Vegetation in the Sagebrush-Grass Area of the Intermountain West," *Journal of Range Management* 28:32-36.

Vale, T.R. 1975b. "Invasion of Sagebrush by White Fir on the East Slope of the Warner Mountains, California," *Great Basin Naturalist* 35:319-324.

Vale, T.R. 1977. "Forest Changes in the Warner Mountains, California," *Annals,* Association of American Geographers 67:28-45.

Vale, T. R. 1978. Proposed Expansion of Redwood National Park, California," *Environmental Conservation* 5:150-151.

Vale, T. R. 1979. "Coulter Pine and Wild Fire on Mount Diablo, California," *Madrono* 26:135-140.

Vale, T. R. 1981. "Tree Invasion of Montane Meadows in Oregon," *American Midland Naturalist* 105:61-69.

Van Derender, T. and W. Spaulding. 1979. "Development of Vegetation and Climate in the Southwestern United States," *Science* 204:701-710.

Van der Zande, A. N. et al. 1980. "The Impact of Roads on the Densities of Four Bird Species in an Open Field Habitat — Evidence of a Long-Distance Effect," *Biological Conservation* 18:299-321.

Vankat, J. L. and J. Major. 1978. "Vegetation Changes in Sequoia National Park, California," *Journal of Biogeography* 5:377-402.

Van Wagner, C.E. 1978. "Age-Class Distribution and the Forest Fire Cycle," *Canadian Journal of Forest Research* 8:220-227.

Van Wagtendonk, J. W. 1975. *Refined Burning Prescriptions for Yosemite National Park.* National Park Service, Occasional Paper Number 2.

Van Zant, K. 1979. "Late Glacial and Postglacial Pollen and Plant Macrofossils From Lake West Okoboji, Northwestern Iowa," *Quaternary Research* 12:358-380.
Vasek, F. C. et al. 1975a. "Effects of Power Transmission Lines on Vegetation of the Mojave Desert," *Madrono* 23:114-130.
Vasek, F. C. et al. 1975b. "Effects of Pipeline Construction on Creosote Bush Scrub Vegetation of the Mojave Desert," *Madrono* 23:1-13.
Veblen, T. T. 1981. "Forest Dynamics in South-Central Chile," *Journal of Biogeography* 8:211-247.
Veblen, T. T. and D. H. Ashton. 1978. "Catastrophic Influences on the Vegetation of the Valdivian Andes, Chile," *Vegetatio* 36:149-167.
Veblen, T. T. and G. H. Stewart. 1980. "Comparison of Forest Structure and Regeneration on Bench and Stewart Islands, New Zealand," *New Zealand Journal of Forestry* 3:50-68.
Vitousek, P. M. et al. 1979. "Nitrate Losses from Disturbed Ecosystems," *Science* 204:469-474.
Vollmer, A. T. et al. 1977. "The Impact of Off-Road Vehicles on a Desert Ecosystem," *Environmental Management* 1:115-129.
Wagener, W. 1961. "Past Fire Incidence in Sierra Nevada Forests," *Journal of Forestry* 59:739-747.
Wagner, W. et al. 1978. "Natural Succession on Strip-Mined Lands in Northwestern New Mexico," *Reclamation Review* 1:67-73.
Wagstaff, F. J. 1980. "Impact of the 1975 Wallsburg Fire on Antelope Bitterbrush," *Great Basin Naturalist* 40:299-302.
Wallmo, O. C. (editor). 1981. *Mule and Black-Tailed Deer of North America.* Lincoln: University of Nebraska Press.
Wallmo, O. C. and J. W. Schoen. 1980. "Response of Deer to Secondary Forest Succession in Southeast Alaska," *Forest Science* 26:448-462.
Ward, R. T. 1956. "The Beech Forests of Wisconsin — Changes in Forest Composition and the Nature of the Beech Border," *Ecology* 37:407-419.
Watt, A. S. 1947. "Pattern and Process in the Plant Community," *Journal of Ecology* 35:1-22.
Weaver, H. 1961. "Ecological Changes in the Ponderosa Pine Forest of Cedar Valley in Southern Washington," *Ecology* 42:416-420.
Weaver, T. and D. Dale. 1978. "Trampling Effects of Hikers, Motorcycles, and Horses in Meadows and Forests," *Journal of Applied Ecology* 15:451-457.
Webb, R. and W. Wilshire. 1980. "Recovery of Soils and Vegetation in a Mojave Desert Ghost Town, Nevada, U.S.A.," *Journal of Arid Environments* 3:291-303.
Webber, P. and J. Ives. 1978. "Damage and Recovery of Tundra Vegetation," *Environmental Conservation* 5:171-182.
Wells, C. G. et al. 1979. *Effects of Fire on Soil.* U.S. Forest Service, General Techical Report WO-7.
Wells, P. 1965. "Scarp Woodlands, Transported Grassland Soils, and Concept of Grassland Climate in the Great Plains Region," *Science* 148:246-249.
Wells, P. 1979. "An Equable Glaciopluvial in the West: Pleniglacial Evidence of Increased Precipitation on a Gradient from the Great Basin to the Sonoran and Chihuahuan Deserts," *Quaternary Research* 12:311-325.
Westmann, W. 1979. "Oxidant Effects on California Coastal Sage Scrub," *Science* 205:1001-1003.
Whipple, S. A. and R. L. Dix. 1979. "Age Structure and Successional Dynamics of a Colorado Subalpine Forest," *American Midland Naturalist* 101:142-158.
Whitcomb, R. F. 1977. "Island Biogeography and Habitat Islands of Eastern Forest," *American Birds* 31:3-5.
White, E. M. and F. F. Riecken. 1955. "Brunizem-Gray Brown Podzolic Soil Biosequences," *Proceedings, Soil Science Society of America* 19:504-509.

White, L. D. and W. S. Terry, 1979. "Creeping Bluestem Response to Prescribe Burning and Grazing in South Florida," *Journal of Range Management* 32:36£ 371.

Whitesell, C. D. 1974. *Planting Trials of 10 Mexican Pine Species in Hawaii.* Pacifi Southwest Forest and Range Experiment Station, Research Paper PSW-103.

Whyte, R. J. and B. W. Cain. 1981. "Wildlife Habitat on Grazed or Ungrazed Sma Pond Shorelines in South Texas," *Journal of Range Management* 34:64-67.

Willard, B. and J. Marr. 1970. "Effects of Human Activities on Alpine Tundra Ecosys tems in Rocky Mountain National Park, Colorado," *Biological Conservatio* 2:257-265.

Willard, B. and J. Marr. 1971. "Recovery of Alpine Tundra Under Protection Afte Damage by Human Activities in the Rocky Mountains of Colorado," *Biologica Conservation* 3:181-190.

Wolff, J. and J. Zasada. 1975. *Red Squirrel Response to Clearcut and Shelterwoo Systems in Interior Alaska.* Pacific Northwest Forest and Range Experimer Station, Research Note PNW-255.

Wood, S. H. 1975. *Holocene Stratigraphy and Chronology of Mountain Meadows Sierra Nevada, California,* unpublished Ph.D. dissertation in geology, Californi Institute of Technology.

Wood, C. and T. Nash. 1976. "Copper Smelter Effluent Effects on Sonoran Dese Vegetation," *Ecology* 57:1311-1316.

Woodward, S. L. and R. D. Ohmart. 1976. "Habitat Use and Fecal Analysis of Fera Burros (*Equus asinus*), Chemehueri Mountains, California, 1974," *Journal c Range Management* 29:482-485.

Woodwell, G. M. 1970. "Effects of Pollution on the Structure and Physiology c Ecosystems," *Science* 168:429-433.

Wright, H. A. 1980. *The Role and Use of Fire in the Semidesert Grass-Shrub Type* Intermountain Forest and Range Experiment Station, General Technical Repoı INT-85.

Wright, H. A. and A. W. Bailey. 1980. *Fire Ecology and Prescribed Burning in th Great Plains – A Research Review.* Intermountain Forest and Range Experimer Station, General Technical Report INT-77.

Wright, H. A. et al. 1979. *The Role and Use of Fire in Sagebrush-Grass and Pinyon Juniper Plant Communities.* Intermountain Forest and Range Experiment Statior General Technical Report INT-58.

Wright, H. E. 1976. "The Dynamic Nature of Holocene Vegetation. *Quaternary Re search* 6:581-596.

Wright, H. E. and M. L. Heinselman. 1973. "Introduction," *Quaternary Researcl* 3:319-328.

Yeaton, R. 1978. "A Cyclical Relationship Between *Larrea tridentata* and *Opuntiǝ leptocaulis* in the Northern Chihuahuan Desert," *Journal of Ecology* 66:651-656.

Young, J. A. and R. A. Evans. 1973. "Downy Brome — Intruder in the Plant Succes sion of Big Sagebrush Communities in the Great Basin," *Journal of Range Man agement* 26:410-415.

Young, J. A. and R. A. Evans. 1978. "Population Dynamics After Wildfires in Sage brush Grasslands," *Journal of Range Management* 31:283-289.

Young, J. A. et al. 1974. "Alien Plants in the Great Basin," *Journal of Range Manage ment* 25:194-201.